COMPUTER ANALYSIS OF VISUAL TEXTURES

THE KLUWER INTERNATIONAL SERIES IN
ENGINEERING AND COMPUTER SCIENCE

ROBOTICS: VISION, MANIPULATION AND SENSORS

Consulting Editor

Takeo Kanade
Carnegie Mellon University

Other books in the series:

Robotic Grasping and Fine Manipulation, M. Cutkosky
ISBN 0-89838-200-9

Shadows and Silhouettes in Computer Vision, S. Shafer
ISBN 0-89838-167-3

Perceptual Organization and Visual Recognition, D. Lowe
ISBN 0-89838-172-X

Robot Dynamics Algorithms, R. Featherstone
ISBN 0-89838-230-0

Three Dimensional Machine Vision, T. Kanade (editor)
ISBN 0-89838-188-6

Kinematic Modeling, Identification and Control of Robot Manipulators, H.W. Stone
ISBN 0-89838-237-8

Object Recognition Using Vision and Touch, P.K. Allen
ISBN 0-89838-245-9

Integration, Coordination and Control of Multi-Sensor Robot Systems,
H.F. Durrant-Whyte, ISBN 0-89838-247-5

Motion Understanding: Robot and Human Vision,
W.N. Martin and J.K. Aggrawal (Editors), ISBN 0-89838-258-0

Bayesian Modeling of Uncertainty in Low-Level Vision, R. Szeliski
ISBN 0-7923-9039-3

Vision and Navigation: The CMU Navlab, C. Thorpe (editor)
ISBN 0-7923-9068-7

Task-Directed Sensor Fusion and Planning: A Computational Approach, G.D. Hager
ISBN 0-7923-9108-X

COMPUTER ANALYSIS OF VISUAL TEXTURES

by

Fumiaki Tomita
Electrotechnical Laboratory
Tsukuba, Japan

and

Saburo Tsuji
Osaka University
Toyonaka, Japan

KLUWER ACADEMIC PUBLISHERS
Boston/Dordrecht/London

Distributors for North America:
Kluwer Academic Publishers
101 Philip Drive
Assinippi Park
Norwell, Massachusetts 02061 USA

Distributors for all other countries:
Kluwer Academic Publishers Group
Distribution Centre
Post Office Box 322
3300 AH Dordrecht, THE NETHERLANDS

Library of Congress Cataloging-in-Publication Data

Tomita, Fumiaki.
    Computer analysis of visual textures / by Fumiaki Tomita and
Saburo Tsuji.
        p.   cm. — (Kluwer international series in engineering and
computer science ; #102)
    Includes bibliographical references and index.
    ISBN 0-7923-9114-4
    1. Computer vision.   2. Visual texture recognition.   3. Artificial
intelligence.   I. Tsuji, Saburo, 1932-   .   II. Title.
III. Series.
TA1632.T66   1990
006.3 ' 7—dc20   90-4674
                                                                    CIP

Contents

Preface

This book presents theories and techniques for perception of textures by computer. Texture is a homogeneous visual pattern that we perceive in surfaces of objects such as textiles, tree barks or stones. Texture analysis is one of the first important steps in computer vision since texture provides important cues to recognize real-world objects.

A major part of the book is devoted to two-dimensional analysis of texture patterns by extracting statistical and structural features. It also deals with the shape-from-texture problem which addresses recovery of the three-dimensional surface shapes based on the geometry of projection of the surface texture to the image plane.

Perception is still largely mysterious. Realizing a computer vision system that can work in the real world requires more research and experiment. Capability of textural perception is a key component. We hope this book will contribute to the advancement of computer vision toward robust, useful systems.

We would like to express our appreciation to Professor Takeo Kanade at Carnegie Mellon University for his encouragement and help in writing this book; to the members of Computer Vision Section at Electrotechnical Laboratory for providing an excellent research environment; and to Carl W. Harris at Kluwer Academic Publishers for his help in preparing the manuscript.

COMPUTER ANALYSIS OF VISUAL TEXTURES

Chapter 1

Introduction

When we observe textiles, tree bark, or stones, we perceive that their surfaces are homogeneous, in spite of fluctuations in brightness or color. Such a homogeneous visual pattern is called *texture*. According to Pickett (1970),

> the basic requirement for an optical pattern to be seen as a texture is that there be a large number of elements (spatial variations in intensity or wave length), each to some degree visible, and, on the whole, densely and evenly arrayed over the field of view.

Since any object has its own specific texture, human beings can use textures as cues to recognize many kinds of objects in the world. For example, we can identify the two textures in Figure 1.1 with straw and lawn. Even if we cannot identify them, we can distinguish them; we can draw a boarderline between them. We also perceive depth or orientation of a three-dimensional surface in the image from projective distortions of textures which cover the surface (Figure 1.2). Endowing machines with these visual abilities is not only an interesting problem in computer science, but also practically important for automatic analysis of aerial photographs or medical images and for mobile robots with vision.

In the field of psychology, researchers have studied texture discrimination by the human visual system. Julesz (1975) conjectured that two textures are not discriminable instantly if their second-order statistics

1

Figure 1.1: Textures of straw and lawn (from Brodats, 1966).

Figure 1.2: Perception of surface orientation (from Gibson, 1950).

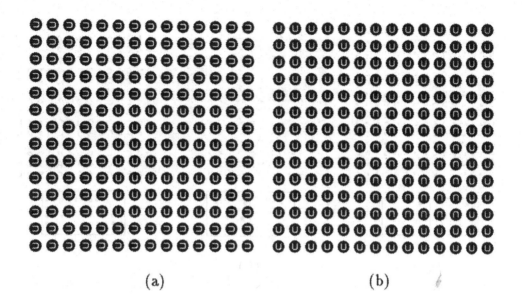

(a)   (b)

Figure 1.3: Texture discrimination: (a) Difference in second-order statistics; (b) Difference in third-order statistics (from Julesz, 1975).

(statistics of pairs of gray values) are identical. For example, we can easily see two different textures in Figure 1.3(a) but it is difficult to distinguish two textures in Figure 1.3(b). The two textures in Figure 1.3(a) differ in their second-order statistics, whereas the two textures in Figure 1.3(b) have the same second-order statistics and differ in third-order statistics (statistics of triplets of gray values).

Julesz's conjecture is valid in the great majority of cases. However, there are clear instances where textures with the same second-order statistics are discriminable on the basis of local geometrical features of (a) colinearity, (b) corner, and (c) closure of texture-element clusters, as shown in Figure 1.4 (Julesz and Caelli, 1979). This implies that there are channels in the human visual system which extract nonlinear local features. These psychological results have affected the works in computer vision.

In computer vision, textures are analyzed on two levels: *statistical* and *structural* (Lipkin and Rosenfeld, 1970). On the statistical level, local features are computed parallelly at each point in a texture image,

3

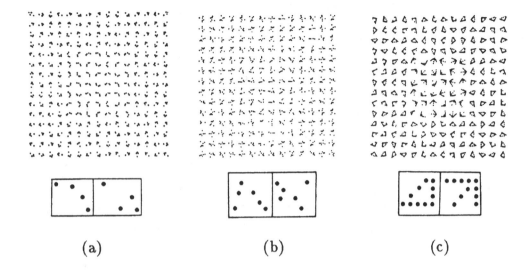

Figure 1.4: Isodipole texture pairs which result in texture discrimination due to the difference in (a) collinearity, (b) corner, and (c) closure properties of the dual micropatterns (from Julesz and Caelli, 1979).

and a set of statistics are derived from the distributions of the local features. The local feature is defined by the combination of intensities at specified positions relative to each point in the image. According to the number of points which define the local feature, statistics are classified into first-order, second-order, and higher-order statistics. Statistics give various measures of texture properties. Statistics need not be only for intensities. Statistics of such local geometrical features as edge, peak and valley, and spot or blob, give measures of specific texture properties. Chapter 2 overviews the measures of texture properties given by statistical texture analysis.

On the structural level, on the other hand, texture is considered to be composed of *texture elements*. Figure 1.5 shows texture elements extracted in a texture image. The properties of texture elements as well as the *placement rules* of texture elements define the texture. The structural analysis is complex in camparison with statistical analysis, and it derives much more detailed information; it is possible to reconstruct the original texture from the resulting description of texture. Chapter

4

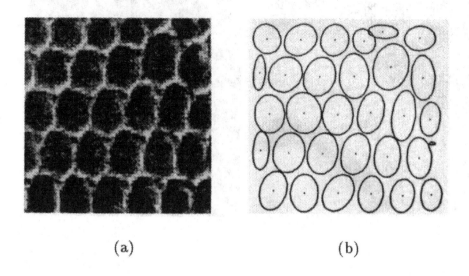

(a)   (b)

Figure 1.5: Structural texture analysis: (a) Reptile skin; (b) Texture elements (approximated by ellipses).

5 presents the algorithms for structural texture analysis.

There are two situations for application of texture analysis. In one situation, there is only one kind of texture in one image. The problem is to classify the image into one of the specified categories of textures. It is called *image classification* and is discussed in Chapter 7.

In the other situation, there are more than one texture in one image. The problem is to extract regions covered with the same textures, or to detect edges between different textures. It is called *image segmentation*. Image segmentation is actually the first step of object recognition, because each region usually corresponds to one individual object, or each edge to the boundary between different objects. Figure 1.6 shows an example of image segmentation. Chapter 3 presents the algorithms for image segmentation.

When texture elements are extracted on the structural level of analysis, clustering the texture elements with similar properties and arranged by the same placement rule results in regions each of which is covered with the same texture. It is called *grouping* in contrast with image segmentation. Figure 1.7 shows an example of grouping. Chapter 6 presents

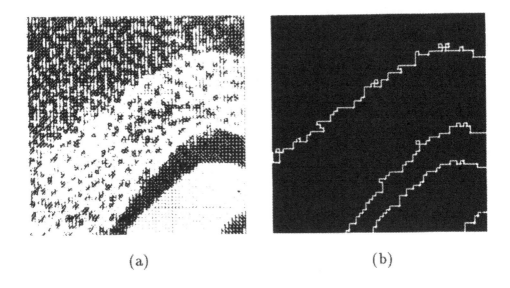

(a)   (b)

Figure 1.6: Image segmentation: (a) Microscopic image of eye ball cells; (b) Boundaries between different textures.

the algorithms for grouping.

Shape of a region extracted in the image is the important cue for recognizing the corresponding object as well as texture. Chapter 4 presents four basic algorithms for shape analysis of a region: the scalar transform of the area which is useful for rough classification of shape, the Fourier transform of the boundary which is good at analyzing curved shape, the medial-axis transformation of the area which is good at analyzing elongate shape, and the segmentation of the boundary which is good at analyzing polygonal shape.

Which level of texture analysis should be used? Suppose a human being sees a newspaper. If he has no interest in reading it, he may regard it as a sheet of paper having almost uniform texture. When his interest in reading is deep, on the other hand, he can recognize each character. Generally speaking, statistical analysis is suitable for fine micro-textures, and structural analysis for coarse macro-textures. Either approach alone cannot deal with all kinds of textures properly. Textures should be analyzed statistically or structurally depending on the nature of each texture. Consider four illustrative patterns of textures

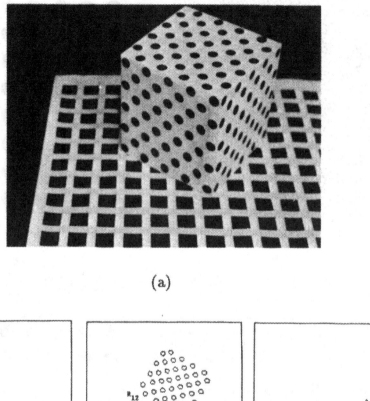

(a)

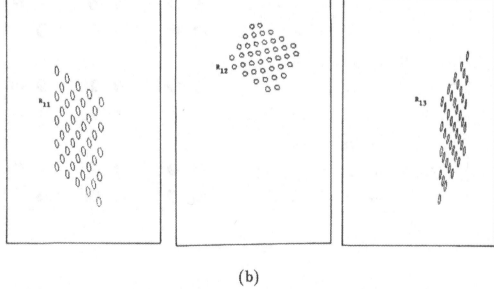

(b)

Figure 1.7: Grouping: (a) A cube on a table; (b) Groups of texture elements on three planes of the cube (from Tsuji and Tomita, 1973).

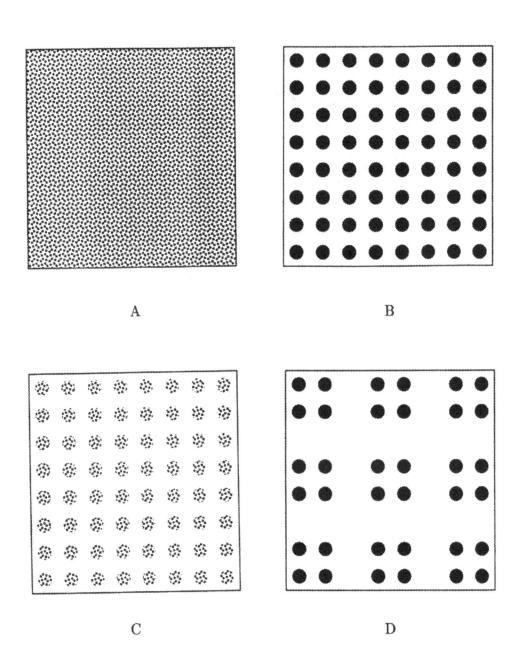

A

B

C

D

Figure 1.8: Four illustrative patterns of textures.

in Figure 1.8. The typical way to analyze each of these patterns is as follows.

Pattern A consists of many small texture elements. It is analyzed statistically without regard to the texture elements. Pattern B consists of large texture elements. It is analyzed structurally based on the texture elements. Pattern C consists of many small texture elements which form local clusters. The clusters are detected statistically by image segmentation without regard to the texture elements, and the pattern is analyzed structurally based on the clusters. Pattern D consists of large texture elements which form local clusters. The clusters are detected structurally by grouping texture elements, and the pattern is analyzed structurally based on the clusters.

Chapter 7 presents this adaptive texture analysis system, and also gives the procedure of not only image classification but also *texture synthesis* as a tool for evaluation, reviewing *generative models* of textures.

The ultimate object of image analysis by computer, which includes texture analysis, image segmentation, shape analysis, and grouping so far discussed, is to automatically recognize objects in the images. The difficulty of recognition is in variety of analysis; different programs are needed to analyze different kinds of images. Generally speaking, the computer must learn about objects in the images before recognition. Learning is currently the main issue in artificial intelligence. The learning strategies are ordered from the easiest one to the most difficult one according to the amount of inference involved: 1) learning by being programmed, 2) learning by being told, 3) learning by seeing samples, and 4) learning by discovery (Winston, 1977).

The first strategy has been the conventional way in which image processing experts have coded a special program whenever a new kind of image is analyzed, developing new algorithms. Now that there are many useful algorithms and even commercialized subroutine libraries are available (Tamura et al., 1983), it is necessary to develop a system with higher learning strategies in which even a user who is not an image processing expert, for example, a physician, can easily create a program which analyzes a specific kind of image, for example, medical images. The current image processors work by commands, the second strategy. However, it is still difficult to use them if the user does't know the spec-

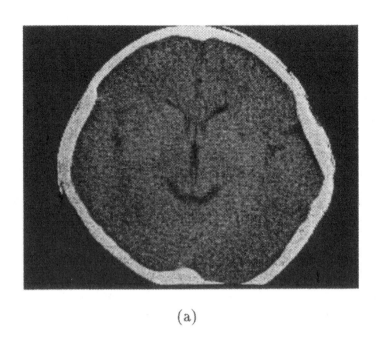

(a)

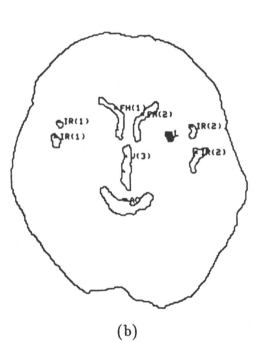

(b)

Figure 1.9: Object recognition: (a) A cranial CT image; (b) Identification of normal cerebrospinal fluid regions and one abnormality (black) (from Tomita, 1988).

10

ifications of programs. The expert systems for image processing teach the user how to select proper programs by production rules (Matsuyama, 1986). However, they still belong to the second strategy.

Chapter 8 presents an object recognition system which supports the user to produce a specific program to recognize specified objects in a specified kind of image with little knowledge of image processing algorithms, by the third strategy. The system is applied to diagnose cranial CT images, as shown in Figure 1.9.

So far we have discussed only two-dimensional features of images. When the image is a projection of a three-dimensional (3D) scene, some 3D forms in the scene can be estimated from 2D features extracted in the image. The role of texture as a basis for the recovery of surface orientation was first investigated, also in the field of psychology, by Gibson (1950). The method to estimate the surface orientation from the apparent texture distortion is called *shape-from-texture*. Texture distortions under perspective projection are caused by the gradient of the surface, the angle between the line of sight and the image plane, and the distance from the view point to the surface. Then, the problem is how to know these factors from the given image. Since the problem is under constraints, some additional assumtions are necessary. Accordingly, various algorithms have been proposed to solve this problem.

Chapter 9 overviews the algorithms for shape-from-texture, classifying them by the surface cue—texture gradient, converging lines, normalized texture property map, or shape distortion, the surface type—planar or curved, a priori knowledge about the original texture, the projection type—orthographic, perspective, or spherical, the level of texture analysis—statistical or structural, the unit of texture analysis, and the property of the unit. The assumption common to all the algorithms is that the surface is smooth and is covered with the homogeneous texture.

Chapter 2

Statistical Texture Analysis

Statistical texture analysis computes local features parallelly at each point in a texture image, and derives a set of statistics from the distributions of the local features. The local feature is defined by the combination of intensities at specified positions relative to each point in the image. According to the number of points which define the local feature, statistics are classified into first-order, second-order, and higher-order statistics. Statistics give various measures of texture properties. Statistics need not be only for intensities. Statistics of such local geometrical features as edge, peak and valley, and spot or blob, give measures of specific texture properties. This chapter overviews the measures of texture properties given by statistical texture analysis.

2.1 First-Order Statistics

2.1.1 Statistical Test

Let $\phi(i)$ $(i = 1, \ldots, n)$ be the number of points whose intensity is i in the image and A be the area of the image (the total number of points in the image). The occurrence probability of intensity i in the image is computed by

$$h(i) = \phi(i)/A \tag{2.1}$$

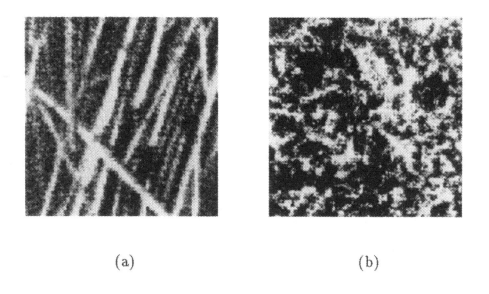

(a)   (b)

Figure 2.1: Digitized images of textures: (a) Straw; (b) Lawn.

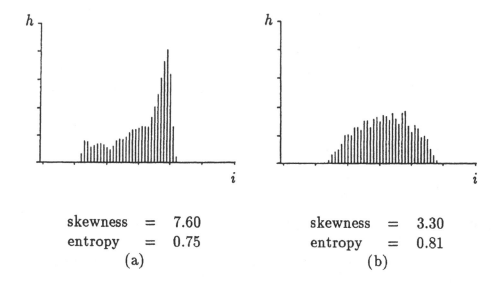

skewness	=	7.60		skewness	=	3.30
entropy	=	0.75		entropy	=	0.81

(a)   (b)

Figure 2.2: Intensity histograms of the images in Figure 2.1.

14

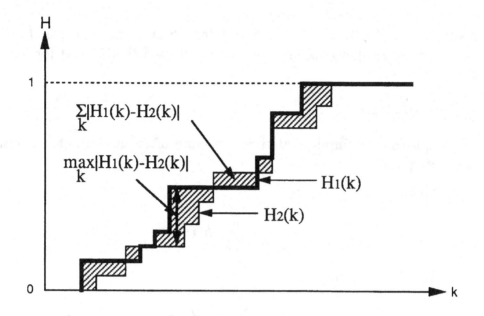

Figure 2.3: Kormogorv-Smilnov test.

The distribution takes the form of a histogram. For example, the intensity histograms of the images in Figure 2.1 are shown in Figure 2.2.

The similarity of two histograms, typically one from the observed data and the other from the reference data, can be compared by the Kolmogorov-Smirnov (KS) test, as shown in Figure 2.3. Let $H(k)$ ($k = 1, \ldots, n$) be the cumulative distribution function derived from $h(i)$ by

$$H(k) = \sum_{i=1}^{k} h(i) \qquad (2.2)$$

When the cumulative distribution functions of two histograms are $H_1(k)$ and $H_2(k)$, the similarity is defined by

$$\max_{k} |H_1(k) - H_2(k)|$$

or

$$\sum_{k} |H_1(k) - H_2(k)|$$

15

Nagao et al. (1976) used the KS test for agricultural land use classification of aerial photographs. Muerle (1970) used the KS test for image segmentation.

2.1.2 Statistics

The following simple statistics are more often used to characterize the histogram.

1. Mean:

$$\sum_{i=1}^{n} ih(i)$$

2. Standard Deviation:

$$\sum_{i=1}^{n} (i - \mu)^2 h(i)$$

3. Third Moment:

$$\sum_{i=1}^{n} (i - \mu)^3 h(i)$$

4. Entropy:

$$-\sum_{i=1}^{n} h(i) \log h(i)$$

where μ is the mean of intensities.

When the problem is to classify texture images, the images are usually normalized to have the same mean and the same standard deviation because these statistics are affected by the image input conditions. The third moment measures the skew of the histogram. When the histogram is symmetrical, the value is 0. When the histogram is skewed left or right, the value is accordingly negative or positive. The entropy measures the uniformity of the histogram. When the distribution is uniform, the entropy takes the maximal value. When there is a dense cluster in the histogram, the value approaches 0. Examples of third moments and entropies are shown below the histograms in Figure 2.2.

16

2.2   Second-Order Statistics

2.2.1   Co-Occurrence Matrix

Let $\delta = (r, \theta)$ denote a vector in the polar coordinates of the image, as shown in Figure 2.4(a). For any such vector, we can compute the joint probability of the pairs of gray levels that occur at pairs of points separated by δ. This joint probability takes the form of an array P_δ, as shown in Figure 2.4(b), where $P_\delta(i,j)$ is the probability of the pair of gray levels (i,j) occurring at separation δ. This array is called the *co-occurrence matrix*.

Finding co-occurrence matrices for all δ requires a prohibitive amount of computation. Haralick, Shanmugam, and Dinstein (1973), who first used co-occurrence matrices to classify terrain in aerial photographs, computed just four co-occurrence matrices for $r = 1$ and $\theta = 0°$, $45°$, $90°$, and $135°$. From each matrix, they computed 14 properties for discriminating between textures. Some of them are as follows.

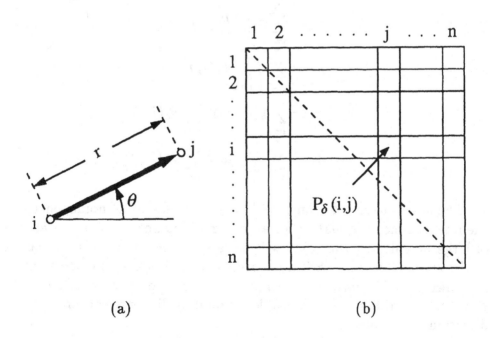

(a)   (b)

Figure 2.4: (a) Displacement $\delta = (r, \theta)$; (b) Co-occurrence matrix.

17

1. Angular Second Moment:

$$\sum_{i=1}^{n}\sum_{j=1}^{n} P_\delta(i,j)^2$$

2. Contrast:

$$\sum_{k=0}^{n-1} k^2 \sum_{\substack{i=1 \\ |i-j|=k}}^{n}\sum_{j=1}^{n} P_\delta(i,j)$$

3. Correlation:

$$\frac{\sum_{i=1}^{n}\sum_{j=1}^{n} i \cdot j P_\delta(i,j) - \mu_x \mu_y}{\sigma_x \sigma_y}$$

where

$$\mu_x = \sum_{i=1}^{n} i \sum_{j=1}^{n} P_\delta(i,j)$$

$$\mu_y = \sum_{j=1}^{n} j \sum_{i=1}^{n} P_\delta(i,j)$$

$$\sigma_x = \sum_{i=1}^{n}(i - \mu_x)^2 \sum_{j=1}^{n} P_\delta(i,j)$$

$$\sigma_y = \sum_{j=1}^{n}(j - \mu_y)^2 \sum_{i=1}^{n} P_\delta(i,j)$$

The angular second-moment (ASM) is a measure of homogeneity of the image. Since the matrix for a homogeneous image has fewer entries of large magnitude, the ASM will be large. The contrast is a measure of the amount of local variations present in the image. The correlation is a measure of linearity in the image. A large correlation value in the direction θ implies a considerable amount of linear structure in that direction in the image.

There are some variations of this method. Since Haralick's properties are redundant, Tou et al. (1977) proposed the use of Karhunen-Loeve

18

(KL) expansion to extract optimal properties from the full property set. Zobrist et al. (1975) used the following distance function to measure the similarity of textures of two regions in a simulation of Beck's psychological experiments.

$$D = c_1 d_1 + \cdots + c_m d_m \tag{2.3}$$

where d_i $(i = 1, \ldots, m)$ is the difference of the ith property between the two regions. The 43 properties were derived mainly from the co-occurrence matrices of $r = 1, 2$, and $\theta = 0°, 45°, 90°, 135°$. Coefficients c_i were determined by the linear programming method. The number of effective properties, whose coefficient is not zero, was only 7. Conners (1979) used the contrast measure in the direction $\theta = 0°, 90°$, as the function of r, to know the horizontal and vertical periodicities of textures, respectively. Deutsch et al. (1972) made direct use of the values in the 2×2 co-occurrence matrices of $r = 2, 4, 8, 16$, and $\theta = 0°, 45°, 90°, 135°$, to analyze binary texture images.

2.2.2 Difference Statistics

The difference statistics are the distribution of probability $P_\delta(k)$ $(k = 0, \ldots, n-1)$ that the gray-level difference is k between the points separated by δ in the image. Difference statistics are a subset of the co-occurrence matrix, and are derived from the matrix by

$$P_\delta(k) = \sum_{\substack{i=1 \\ |i-j|=k}}^{n} \sum_{j=1}^{n} P_\delta(i, j) \tag{2.4}$$

Weszka et al. (1976a and 1976b) computed the following properties from the distribution.

1. Angular Second Moment:

$$\sum_{k=0}^{n-1} P_\delta(k)^2$$

2. Contrast:

$$\sum_{k=0}^{n-1} k^2 P_\delta(k)$$

3. Entropy:

$$-\sum_{k=0}^{n-1} P_\delta(k) \log P_\delta(k)$$

4. Mean:

$$\sum_{k=0}^{n-1} k P_\delta(k)$$

2.2.3 Fourier Power Spectrum

Textures can be analyzed in the frequency domain by Fourier transform of the image. Let $f(x, y)$ be a texture image of size $I \times J$. The two-dimensional discrete Fourier transform is defined by

$$F(u, v) = \sum_{x=0}^{I-1} \sum_{y=0}^{J-1} f(x, y) \exp^{-2\pi\sqrt{-1}(ux/I + vy/J)} \tag{2.5}$$

The power spectrum which represents the strength of each spatial frequency is obtained by

$$P(u, v) = |F(u, v)|^2 \tag{2.6}$$

Bajcsy (1973) computed the following distributions from the power spectrum $P(r, \theta)$ in the polar coordinates.

$$P(r) = 2 \sum_{\theta=0}^{\pi} P(r, \theta)$$
$$P(\theta) = \sum_{r=0}^{n/2} P(r, \theta) \tag{2.7}$$

20

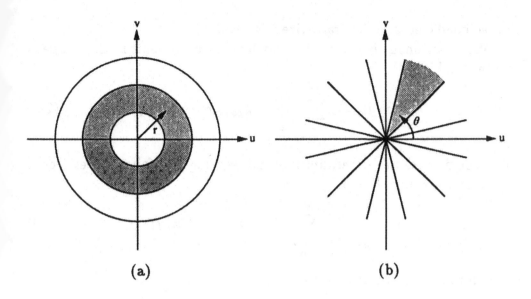

Figure 2.5: Fourier power spectrum: (a) Ring filter; (b) Wedge filter.

$P(r)$ and $P(\theta)$ are the sums of the powers in ring and wedge regions of the power spectrum space, respectively, as shown in Figure 2.5. The peak in $P(r)$ indicates the size of the dominant texture elements or the texture coarseness. The peak in $P(\theta)$ indicates the directivity in that direction in the texture.

2.2.4 Autoregression Model

In the field of signal processing, time series analysis is a well-known method for prediction of a one-dimensional signal from the past signal values. The method is extended to image processing by changing the concept of one-dimensional time to two-dimensional space (Deguchi and Morishita, 1978). Let $f(x, y)$ be the gray level at point (x, y) in the image and be generated by a linear combination of the gray levels in its neighborhood with white noise $w(x, y)$ added:

$$f(x, y) = \sum_{\substack{p=-M \\ (p,q)\neq(0,0)}}^{M} \sum_{q=-N}^{N} a_{pq} f(x - p, y - q) + w(x, y) \qquad (2.8)$$

These coefficients a_{pq} characterize the texture.

When an image is given, the gray level at each point in the image is estimated by

$$\hat{f}(x,y) = \sum_{\substack{p=-M \\ (p.q)\neq(0,0)}}^{M} \sum_{q=-N}^{N} a_{pq} f(x-p, y-q) \qquad (2.9)$$

The coefficients are determined by minimizing the following expected value of the squared error,

$$S = \sum_{\substack{p=-M \\ (p.q)\neq(0,0)}}^{M} \sum_{q=-N}^{N} E[(\hat{f}(x,y) - f(x,y))^2] \qquad (2.10)$$

The minimum value of S is given by

$$S_{min} = C(0,0) - \sum_{p=-M}^{M} \sum_{q=-N}^{N} a_{pq} C(p,q) \qquad (2.11)$$

where $C(p,q)$ is the autocorrelation function of the image.

The window size $(2M+1) \times (2N+1)$ also depends on the texture and must first be determined. The window size is determined by the *Final Prediction Error* (FPE) criterion. Let the gray level at the current point (x,y) be estimated by the points on the horizontal and vertical lines which pass through the current point:

$$\hat{f}(x,y) = \sum_{p=1}^{M} a_p^{(M)} f(x-p, y)$$

$$\hat{f}(x,y) = \sum_{q=1}^{N} a_q^{(N)} f(x, y-q) \qquad (2.12)$$

The optimum coefficients $a_p^{(M)}$ and $a_q^{(N)}$ and the minimum estimation error $S_{min}^{(M)}$ and $S_{min}^{(N)}$ can also be obtained by the least squares method. When the size of the image is $I \times J$, the optimum values of M and N are those which minimize the following cost functions.

$$FPE(M) = \frac{I+M+1}{I-M-1} S_{min}^{(M)}$$

$$FPE(N) = \frac{J+N+1}{J-N-1} S_{min}^{(N)} \qquad (2.13)$$

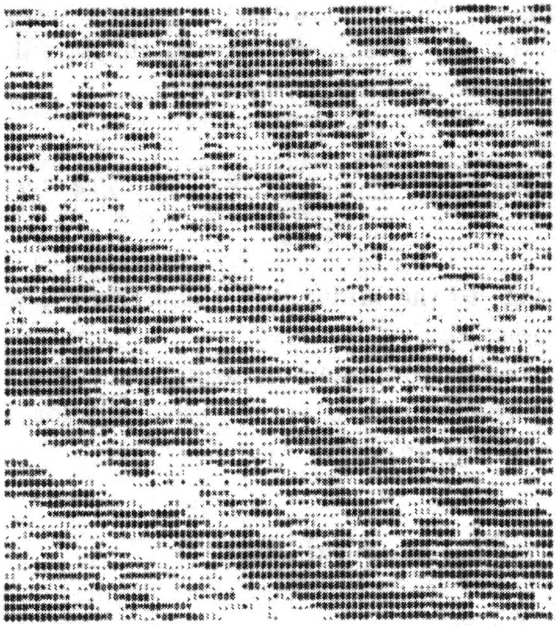

(a)

			-p	0	p		
-q	-0.0002	-0.0292	0.0540	0.0005	0.0013	-0.0048	0.0015
	0.0152	-0.0007	-0.0087	-0.0956	-0.0029	0.0102	0.0021
	-0.0084	-0.0180	-0.0053	0.5201	-0.0862	-0.0045	0.0005
0	0.0046	0.0076	0.1428		0.1459	0.0165	-0.0095
	-0.0111	0.0020	-0.0816	0.5192	-0.0067	-0.0182	-0.0067
q	0.0038	0.0089	-0.0030	-0.0963	-0.0081	0.0001	0.0142
	-0.0032	-0.0021	0.0025	0.0008	0.0539	-0.0290	0.0045

(b)

Figure 2.6: Autoregression model: (a) Texture image; (b) Coefficients (from Deguchi and Morishita, 1978).

Examples of these coefficients are shown in Figure 2.6. These coefficients are used to determine whether a unknown texture $g(x, y)$ belongs to the same class as $f(x, y)$. The error between the actual value and the estimated value is given for all (x, y) by

$$e(x, y) = g(x, y) - \sum_{\substack{p=-M \\ (p.q) \neq (0,0)}}^{M} \sum_{q=-N}^{N} a_{pq} g(x - p, y - q) \qquad (2.14)$$

If the texture belongs to the same class, the errors are white noise and the deviation is about S_{min}.

Deguchi et al. applied this method to image segmentation. But, it has been proved that the method can be used only for texture classification but for image segmentation.

2.3 Higher-Order Statistics

2.3.1 Interval Cover

Read et al. (1972) used a filter whose size is $m = 3 \times 2$ to classify chromatins and artifacts in microscopic pictures of cells. A pattern $e = (x_1, x_2, \ldots, x_m)$ is called an *event*. Let E^1 and E^0 denote the sets of events in the training images which belong to two categories T^1 and T^0, respectively. When the likelihood ratio of an event e is

$$LR(e) = \frac{P(e \mid T^1)}{P(e \mid T^0)} \qquad (2.15)$$

the output of the filter is

$$\Phi_R(e) = 1 \quad \text{iff } e \in R \qquad (2.16)$$

where

$$\begin{aligned}
&F^{1\beta} \subseteq R \subseteq F^{1\beta} \cup F^{*\beta} \text{ and } R \cap F^{0\beta} = \phi \\
&F^{1\beta} = \{e \mid e \subseteq E^1 \cup E^0 \text{ and } LR(e) > \beta\} \\
&F^{0\beta} = \{e \mid e \subseteq E^1 \cup E^0 \text{ and } LR(e) \leq \beta\} \\
&F^{*\beta} = \{e \mid e \subseteq \overline{E^1 \cup E^0}\} \text{ (don't care)}
\end{aligned} \qquad (2.17)$$

R is the minimal interval cover which includes $F^{1\beta}$ defined by switching theory.

2.3.2 Run Length

The run-length matrix $P_\theta(i,j)$ ($i = 1,\ldots,m$, $j = 1,\ldots,n$) represents the frequency that j points with gray level i continue in the direction θ. Galloway (1975) computed the following five properties from
run-length matrices of $\theta = 0°$, $45°$, $90°$, and $135°$, for terrain classification of aerial photographs.

1. Short Runs Emphasis:

$$\sum_{i=1}^{n}\sum_{j=1}^{l}\frac{P_\theta(i,j)}{j^2} \bigg/ \sum_{i=1}^{n}\sum_{j=1}^{l} P_\theta(i,j)$$

2. Long Runs Emphasis:

$$\sum_{i=1}^{n}\sum_{j=1}^{l} j^2 P_\theta(i,j) \bigg/ \sum_{i=1}^{n}\sum_{j=1}^{l} P_\theta(i,j)$$

3. Gray Level Nonuniformity:

$$\sum_{i=1}^{n}\left\{\sum_{j=1}^{l} P_\theta(i,j)\right\}^2 \bigg/ \sum_{i=1}^{n}\sum_{j=1}^{l} P_\theta(i,j)$$

4. Run Length Nonuniformity:

$$\sum_{j=1}^{l}\left\{\sum_{i=1}^{n} P_\theta(i,j)\right\}^2 \bigg/ \sum_{i=1}^{n}\sum_{j=1}^{l} P_\theta(i,j)$$

5. Run Percentage:

$$\sum_{i=1}^{n}\sum_{j=1}^{l} P_\theta(i,j) \bigg/ A$$

where A is the area of the image.

2.4 Local Geometrical Features

2.4.1 Edge

Edges are located where the intensity abruptly changes. Let $f(x, y)$ denote the value at point (x, y) in the image. The strength and the direction of the edge element at (x, y) are computed as

$$|e(x,y)| = \sqrt{f_x(x,y)^2 + f_y(x,y)^2}$$
$$\angle e(x,y) = \tan^{-1} \frac{f_y(x,y)}{f_x(x,y)} + \frac{\pi}{2} \quad (2.18)$$

where

$$f_x(x,y) = f(x+1,y) - f(x-1,y)$$
$$f_y(x,y) = f(x,y+1) - f(x,y-1) \quad (2.19)$$

For example, Figure 2.7 shows the strong edge elements after the weak edge suppression and the nonmaximal edge suppression detected in the images in Figure 2.1. The following texture properties are derived from first-oder statistics of the distributions of the edge elements.

1. Coarseness: The density of edge elements is a measure of texture coarseness. The finer the texture, the higher the edge density.

2. Contrast: The mean of edge strengths is a measure of contrast.

3. Randomness: The entropy of the histogram of edge strengths is a measure of randomness.

4. Directivity: The directivity is detected from the histogram of edge directions. The entropy of the histogram gives a rough measure of directivity. The directions of directivity are given by detecting clusters in the histogram. If $2N$ ($N > 0$) clusters are detected, they indicate the directivity in N directions. If the distribution is uniform, however, it indicates no directivity in the texture. Figure 2.8 shows the edge-direction histograms of the edges in Figure 2.7. The straw has the directivity in one direction, whereas the grass has no directivity.

The following structural texture properties are extracted from the second-order statistics of edge directions.

26

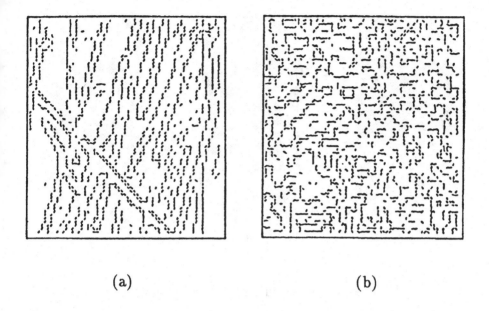

(a)   (b)

Figure 2.7: Edge elements of the images in Figure 2.1.

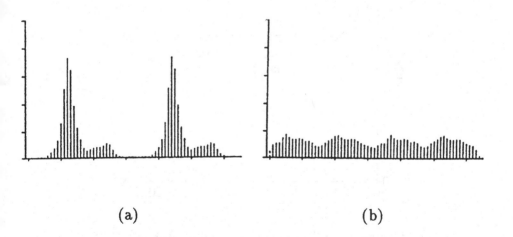

(a)   (b)

Figure 2.8: Edge direction histograms.

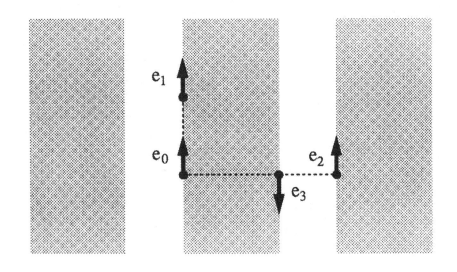

Figure 2.9: Co-occurrence of edge elements.

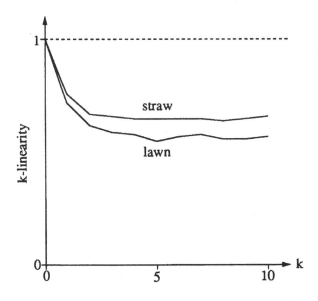

Figure 2.10: k-linearities of edge elements.

28

1. Linearity: The co-occurrence probability of two edge elements with the same edge direction separated by distance k in the edge direction (edge elements e_0 and e_1 in Figure 2.9) indicates the linearity of texture. Figure 2.10 shows the k-linearities of the edges in Figure 2.7. Straw has higher linearity than lawn.

2. Periodicity: The co-occurrence probability of two edge elements with the same direction separated by distnace k along the direction perpendicular to the edge direction (edge elements e_0 and e_2 in Figure 2.9) indicates the *periodicity* of the texture.

3. Size: The co-occurrence probability of two edge elements with the opposite direction separated by distnace k along the direction perpendicular to the edge direction (edge elements e_0 and e_3 in Figure 2.9) represents the *size* of the texture elements.

2.4.2 Peak and Valley

Mitchell et al. (1977) computed the number of local extrema (peaks and valleys) of various sizes by scanning a texture image horizontally and vertically. The following *gear backlash smoothing* is used to detect the local extrema of size T, as shown in Figure 2.11. Let x_k be the value of the kth point and y_k be its value after smoothing. Starting from $y_1 = x_1$, the succeeding y_k is defined as follows.

if	then
$y_k < x_{k+1} - \frac{T}{2}$	$y_{k+1} = x_{k+1} - \frac{T}{2}$
$x_{k+1} - \frac{T}{2} < y_k < x_{k+1} + \frac{T}{2}$	$y_{k+1} = y_k$
$x_{k+1} + \frac{T}{2} < y_k$	$y_{k+1} = x_{k+1} + \frac{T}{2}$

$$(2.20)$$

The peaks and the valleys are defined as the points with the local maximum and the local minimum of y values, respectively. The numbers of local extrema of various sizes characterize textures.

Ehrick et al. (1978) proposed other algorithms to detect peaks in an image. They used the histograms of absolute height, relative height, and width of each size of peaks.

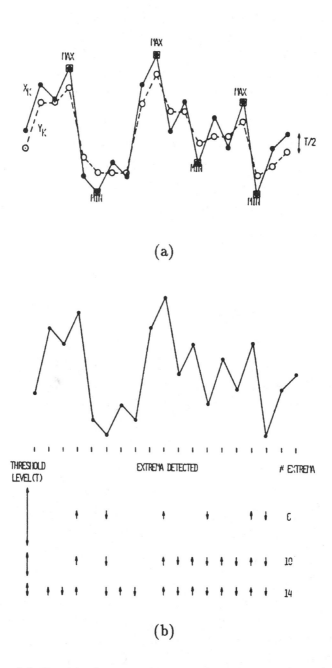

(a)

(b)

Figure 2.11: (a) Backlash smoothing and extrema identification for a threshold distance of T; (b) Extrema detected for three different thresholds (from Mitchell, Myer, and Boyne, 1977).

2.4.3　Spot

Rosenfeld et al. (1971) proposed a spot (blob) detector of size (M, N) $(M > N)$ at a point (x, y) in a texture image:

$$
s(x, y) = \left| \frac{1}{(2N+1)^2} \sum_{u=x-N}^{x+N} \sum_{v=y-N}^{y+N} f(u, v) \right.
$$
$$
\left. - \frac{1}{(2M+1)^2} \sum_{u=x-M}^{x+M} \sum_{v=y-M}^{y+M} f(u, v) \right| \tag{2.21}
$$

Hayes et al. (1974) defined the spot size as k when the value of a spot detector $(2^k, 2^{k-1})$ $(k = 2, \ldots, 5)$ is the maximum. Any spot whose value is not the maximal in its receptive field is deleted (nonmaximal suppression). The average of spot sizes correlates with coarseness of textures. Davis et al. (1979) used the regularity of arrangements of spots in the analysis of aerial photographs of spot-like orchards.

2.4.4　Primal Sketch

Marr (1976) proposed the symbolic representation of an image which is called the *primal sketch*. The image is correlated with edge and bar masks of various sizes and directions to detect primitives like edges, lines, and blobs. Each primitive has attributes such as orientation, size, contrast, position, and termination points. Figure 2.12 is an example of the primal sketch. The following statistics are computed for texture classification.

1. The total amount of contour, and the number of blobs, at different contrasts and intensities.

2. Orientations: the total number of elements at each orientation, and the total contour length at each orientation.

3. Size: distribution of the size parameters defined in the primal sketch.

4. Contrast: distribution of the contrast of items in the primal sketch.

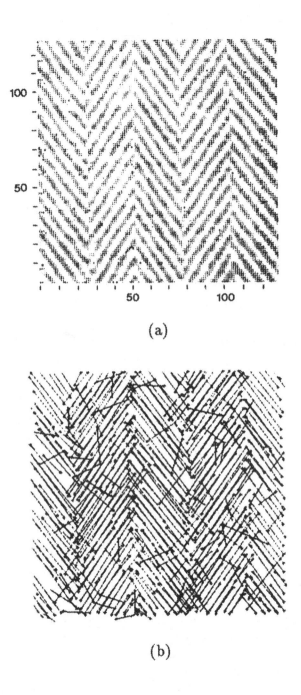

Figure 2.12: (a) Digitized image of texture; (b) Primal sketch (from Marr, 1976).

32

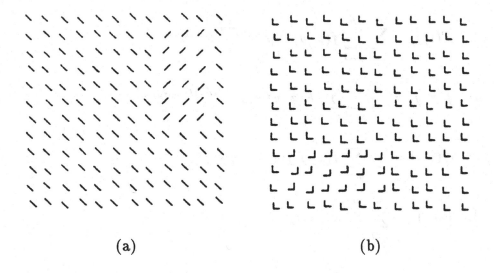

(a)   (b)

Figure 2.13: (a) Textures with different orientation of lines; (b) Textures with the same length and orientation of lines (from Shatz, 1977).

5. Spatial Density: spatial density of place-tokens defined in the different possible ways, measured using a small selection of neighborhood sizes.

Length and orientation of lines are useful properties to discriminate textures such as Figure 2.13(a). However, this is not enough. Though textures such as Figure 2.13(b) have the same length and orientation of lines, they are discriminable. Schatz (1977) proposed the use of length and orientation, not only of actual lines but also local virtual lines between terminators of actual lines. This captures almost all second-order statistics in the point-and-line textures.

2.5 Summary

We have overviewed the statistical methods to characterize textures. It is important to know the ability of each method. As for the amount of information, the methods based on second-order statistics are interrelated, as shown in Figure 2.14. The mathematical relationships between them are summarized in the following.

1: $F(i,j) = \sum\limits_{x=1}^{n} \sum\limits_{y=1}^{n} f(x,y) \exp^{-2\pi\sqrt{-1}(ix+iy)/n}$

2: $f(x,y) = \dfrac{1}{n^2} \sum\limits_{x=1}^{n} \sum\limits_{y=1}^{n} F(x,y) \exp^{-2\pi\sqrt{-1}(ix+iy)/n}$

3: $C(i,j) = \sum\limits_{x=1}^{n} \sum\limits_{y=1}^{n} f(x,y) f(x+i, y+j)$

4: $P_\delta(k) = \sum\limits_{\substack{i=1 \\ |i-j|=k}}^{n} \sum\limits_{j=1}^{n} P_\delta(i,j)$

5: $C(i,j) = n^2 \sum\limits_{u=1}^{n} \sum\limits_{v=1}^{n} P_\delta(u,v) \cdot u \cdot v$

6: $P(u,v) = |F(u,v)|^2$

7: $P(i,j) = \sum\limits_{u=1}^{n} \sum\limits_{v=1}^{n} C(u,v) \exp^{-2\pi\sqrt{-1}(iu+iv)/n}$

8: $C(i,j) = \dfrac{1}{n^2} \sum\limits_{u=0}^{n-1} \sum\limits_{v=0}^{n-1} P(u,v) \exp^{-2\pi\sqrt{-1}(iu+iv)/n}$

9: Solution of $\sum\limits_{\substack{p=-M \\ (p,q)\neq(0,0)}}^{M} \sum\limits_{q=-N}^{N} a_{pq} C(p-k, q-l) = C(k,l)$

10: $P(i,j) = \dfrac{S}{1 - \left| \sum\limits_{p=-M}^{M} \sum\limits_{q=-N}^{N} a_{pq} \exp^{-2\pi\sqrt{-1}(ip/M+iq/N)} \right|^2}$

Co-occurrence matrices for all δ represent all second-order statistics. A co-occurrence matrix for a given δ is a subset of the second-order statistics. The difference statistics for the same δ are a subset of the

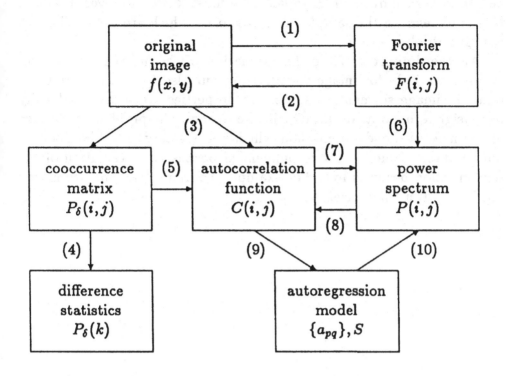

Figure 2.14: Interrelation between second-order statistics.

subset. The Fourier power spectrum, the autoregression model, and the autocorrelation functions are all, theoretically, the same subset of the second-order statistics. Each of them can be mathematically derived from another.

Higher-order statistics do not necessarily give better results in texture discrimination than lower-order statistics. One reason is that they do not cover important information in lower-order statistics. Another reason is that there is little useful information in higher-order statistics for texture discrimination as Julesz (1979) conjectured.

We can easily evaluate each method by testing the correctness of classication of given textures. For example, Weszka et al. (1976a and 1976b) used co-occurrence matrices, difference statistics, run lengths, and the Fourier power spectra to classify terrains in aerial photographs and to inspect industrial materials. We must know, however, that the results depend on the samples and categories which are used. They do not give absolute evaluations.

Another important factor for evaluation is whether the method can be used not only for image classification but also for image segmentation. In image segmentation, texture properties are measured locally in a neighborhood at each point in the image. The problem is the size of the neighborhood. A method which needs a large neighborhood to obtain stable values is not recommended because the resolution of the segmentation becomes low. For example, Fourier power spectrum may not be good in this sense.

Chapter 3

Image Segmentation

When there are more than one texture in one image, the procedure to extract regions covered with the same textures or to detect edges between different textures is called *image segmentation*. Image segmentation is actually the first step of object recognition, because each region usually corresponds to one individual object, or each edge to the boundary between different objects. This chapter presents the algorithms for image segmentation.

3.1 Edge Detection

3.1.1 Intensity Edges

Intensity edges are located where the intensity abruptly changes. Let $f(x, y)$ denote the intensity at point (x, y) in the image. The strength and the direction of the intensity gradient at (x, y) are defined by

$$|e(x, y)| = \sqrt{f_x(x, y)^2 + f_y(x, y)^2}$$

$$\angle e(x, y) = \tan^{-1} \frac{f_y(x, y)}{f_x(x, y)} + \frac{\pi}{2}$$

(3.1)

where f_x and f_y are the first derivatives in x and y directions, respectively. Many operators have been proposed for them. The simplest one is defined by

$$f_x(x, y) = f(x + 1, y) - f(x - 1, y)$$

$$f_y(x, y) = f(x, y + 1) - f(x, y - 1)$$

(3.2)

Since there are tiny fluctuations of intensities as well as noises in the image, it is usual that many noisy weak edges are detected together with real strong edges. Weak edge elements, which have less strength than a given threshold t_1, are suppressed (weak edge suppression).

The profile of the intensity gradient across the real edge is usually bell-shaped because of noise and blur. One method to overcome this is to suppress locally nonmaximal edge elements. Edge elements, which have less strength than the two adjacent edge elements in the direction perpendicular to the edge direction, are suppressed (nonmaximal edge suppression). As a result, we can have only strong edges of one pixel wide, as shown in Figure 3.2(b).

The alternative is to use the second derivative. The edge is detected at the position called *zero-crossing* where the second derivative changes its sign. The zero-crossing corresponds to the local maximum of the first derivative. The sum of the second derivatives f_{xx} and f_{yy} is called Laplacian and some operators have been proposed for it. Recently, circulary symmetric Laplacian called DOG (Difference Of Gaussian), which Marr and Hildreth (1980) proposed from the psychophysical point of view, is often used.

The weak edge suppression suppresses not only noises but also real weak edge elements. As a result, strong but incomplete edges are detected which do not close regions. The method to overcome this is to extend the incomplete edges as follows.

1. Look for a terminal edge element of an incomplete edge by raster scanning the image. It is the starting point of extension.

2. Pick up three adjacent edge elements which are almost in the direction of the current edge element e_0 (edge elements e_5, e_6, and e_7 in Figure 3.1).

3. Of the three edge elements, the one that maximizes the following function is selected.

$$f(e_0, e_i) = |e_i| \cos(\angle e_i - \angle e_0)$$

38

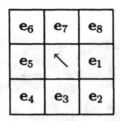

Figure 3.1: 3×3 window for edge extension (the arrow signifies the direction of the central edge point e_0).

This function gives a large value when the strength of the edge element e_i is large and the direction is the same as that of the current edge element e_0.

4. If the selected edge element is a suppresseded edge element, revive it, move the current position to it , and go to step 2.

5. If the selected edge element is an alive (not suppressed) edge element, stop the current extension and return to the starting point.

The same extension is performed backward by changing step 2 with step 2'.

2'. Pick up three adjacent edge elements which are almost in the direction *opposite* to the current edge direction (edge elements e_1, e_2, and e_3 in Figure 3.1).

After both forward and backward extension, resume the raster scanning at step 1 and repeat extension. Figure 3.2(c) shows the result of edge extension.

It may be necessary to check at step 4 if a number of *very* weak edge elements, of which the strength is below a given threshold t_2 ($< t_1$), has been revived successively. If so, the extending edge is a crack (decaying edge) and the extension must be stopped.

3.1.2 Texture Edges

On the analogy of intensity edges, texture edge is located where the texture abruptly changes. Let $A(x, y)$ denote a texture property

39

(a)

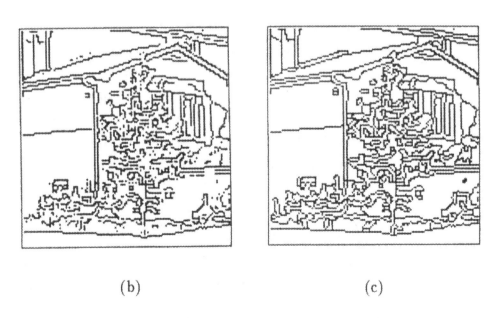

(b)   (c)

Figure 3.2: Edge detection: (a) Input image; (b) Strong edge elements; (c) Result of edge extension.

40

computed in the $n \times n$ neighborhood at point (x, y) in the image. The strength and the direction of the texture gradient at (x, y) are defined by

$$|e^*(x,y)| = \sqrt{A_x(x,y)^2 + A_y(x,y)^2}$$

$$\angle e^*(x,y) = \tan^{-1}\frac{A_y(x,y)}{A_x(x,y)} + \frac{\pi}{2}$$

(3.3)

where A_x and A_y are the first derivatives of the texture property in x and y directions, respectively:

$$A_x(x,y) = A(x + \frac{n}{2}, y) - A(x - \frac{n}{2}, y)$$

$$A_y(x,y) = A(x, y + \frac{n}{2}) - A(x, y - \frac{n}{2})$$

(3.4)

The weak edge suppression and the nonmaximal edge suppression detects strong txeture edges of one pixel wide. The nonmaximal suppression, in this case, compares each edge element with $\frac{n}{2}$ edge elements in the both direction perpendicular to the edge direction and suppresses the edge element if it is not the maximum of them.

Since multiple texture properties are usually computed, the edge detection is recursively applied to each of texture properties to detect multiple texture edges. Another approach is to compute the linear combination of the gradients of the multiple texture properties (Thompson, 1977).

The problem in detecting texture edges is how to decide the size of edge detectors, i.e., the size of neighborhoods in which local texture properties are computed. If the size of the edge detector is too small, the output fluctuates due to local intensity variation, and if it is too large, the real edges are blurred. Rosenfeld et al. (1971; 1972) proposed variable-sized edge detectors to get the best edges.

At each point in the image, the first derivatives in the four directions $(0°, 45°, 90°,$ and $135°)$ are computed using the neighborhood of size $2^k \times 2^k$ $(k = 0, \ldots, L)$. The maximal value D_k of the four derivatives is selected for each size k. Then, the best value of k is determined by the following criterion.

$$D_L < \lambda D_{L-1} < \cdots < \lambda^{L-k} D_k \quad (\lambda = 0.75)$$

(3.5)

41

and

$$D_k \geq \lambda D_{k-1} \qquad (3.6)$$

Applying the variable-sized edge detectors to the image in Figure 3.3(a) gives the edge image in Figure 3.3(b). The nonmaximal edge suppression, of which the scope varies according to the size of 2^k at each point in the image, detects the texture edges of one pixel wide, as shown in Figure 3.3(c).

3.2 Region Extraction

3.2.1 Uniform Regions

There are two basic methods for extracting uniform intensity regions from the image: *region thresholding* (or region splitting) and *region merging* (or region growing). In the thresholding method, the intensity histogram of the image is computed. If there are some clusters in the histogram, the image is segmented into regions by assigning each point with the label of the cluster it belongs to. The thresholding method, however, is too global; even though there are visually distinctive regions in the image, it is usual that separable clusters do not appear in the intensity histogram because of noise and blur. In the merging method, on the other hand, adjacent points in the image are merged into regions if they have similar intensities. The merging method, however, is too local to evaluate the limit of the similarity of intensity.

The *merging-and-thresholding method*, which combines the two methods, offsets the defects of each; the merging method is repeated until separable clusters appear in the intensity histogram, and then the thresholding method is applied. The procedure of extracting regions from the input image I_0 is as follows.

1. Set $k=1$.

2. Generate an image I_k by merging adjacent points whose intensity difference is less than k into regions and averaging the intensities in the regions.

3. Compute the intensity histogram H_k of the image I_k.

42

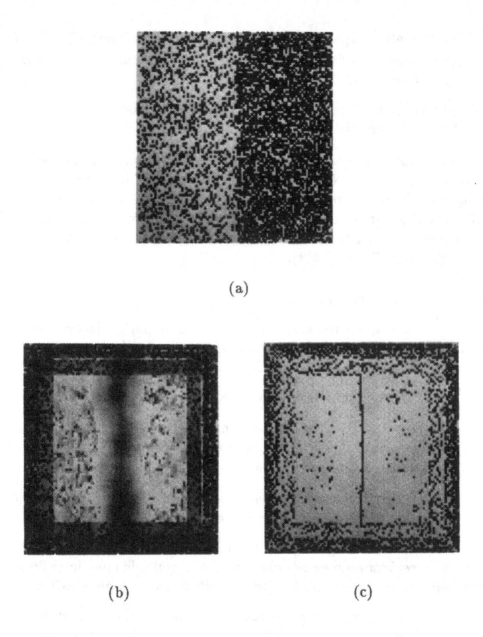

Figure 3.3: Texture edge detection: (a) Input image; (b) Edge image; (c) Nonmaximal edge suppression.

4. Test whether separable clusters exist in the histogram H_k. If so, go to step 5, or else set $k=k+1$ (the increment may be more than 1) and go to step 2.

5. Threshold the image I_0 by the intensities between the clusters in the histogram H_k in order to extract regions.

For example, the image in Figure 3.4(a) has the intensity histogram H_1 in Figure 3.4(b). No separable cluster exists in it. The intensity histograms change by repeating the merging process. Merging by $k=5$ makes two separable clusters appear in the histogram H_5, as shown in Figure 3.4(c). Then, the image is segmented into regions by thresholding, as shown in Figure 3.4(d).

3.2.2 Texture Regions

When the image is represented by multiple properties (color and/or texture), the basic approach for extracting regions with similar properties from the image is as follows (Coleman and Andrews; 1979).

1. Compute the multiple properties at each point of the image.

2. Map every point to the multidimensional property space, and classify the points by clustering, a common technique in the field of statistical pattern recognition.

3. Assign each point with the label of the cluster it belongs to, and merge the adjacent points with the identical labels into regions.

Instead of clustering in the multidimensional space, the following *recursive thresholding method*, which uses the multiple one-dimensional histograms, is more commonly used (Tomita 1973; Ohlander 1979).

1. Select the unsegmented region in the image. Initially, it is the entire image.

2. Compute the histograms for all the multiple properties for the portion of the image which is contained in the region.

3. Select the property that shows the best separation of peaks (clusters) in the histogram.

44

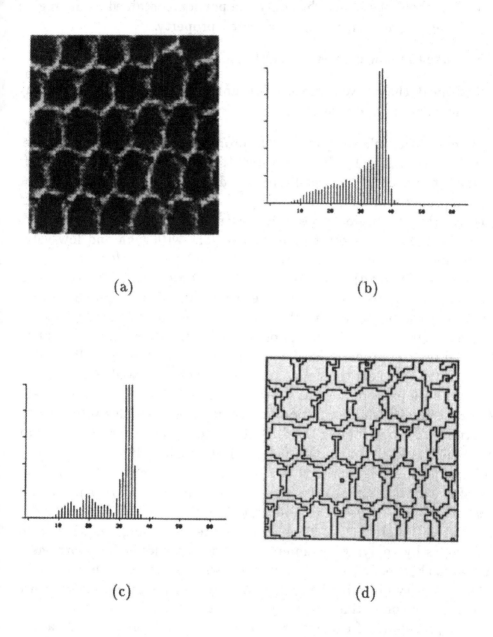

Figure 3.4: Merging-and-thresholding: (a) Reptile skin; (b) Intensity histogram H_1; (c) Intensity histogram H_5; (d) Boundaries of regions extracted by thresholding (from Tomita, 1981).

4. Threshold the image, but only the portion contained in the region being segmented, using the selected property.

5. Make the new connected (sub)regions.

6. Repeat the above segmentation until no peak is detected in any histograms in any regions.

There is the following inevitable problem in extracting texture regions from the image by either method. If the image contains the regions of which the property values lie in two disjoint ranges, computing the texture property in a fixed neighborhood at each point of the image can separate the two region types. However, this method does not work if there are three or more region, and regions with high and low value ranges are adjacent one another; false areas of a region R appear along boundaries of two adjacent regions if the property values of the two regions R_1 and R_2 are higher and lower than that of R. Figure 3.5 shows the result of applying this method to an image with several regions.

The neighborhood $N_{2n}(x, y)$ of the point (x, y) used for computing the local texture property has $2n \times 2n$ points, and its center is located at (x, y). If (x, y) is near the boundary of two regions R_1 and R_2, $N_{2n}(x, y)$ contains both parts of them as shown in Figure 3.6(a). Then the value in $N_{2n}(x, y)$ gives false information for segmenting the image into regions.

An idea for overcoming the difficulty is to search around (x, y) for an area which is likely to contain points of one region. If the boundary in and around the neighborhood is a line or a smooth curve, then at least one of five areas $N_{2n}(x, y)$, $N_{2n}(x + n, y + n)$, $N_{2n}(x - n, y + n)$, $N_{2n}(x - n, y - n)$, $N_{2n}(x + n, y - n)$ around (x, y) is entirely included in one region as shown in Figure 3.6(a). Thus we calculate nonhomogeneity indices of these areas by applying a gradient operator to values in four portions of each area (Figure 3.6(b)), and select one with the least nonhomogeneity index for every point in the image. A better image for segmentation is generated by computing texture properties in the selected areas.

The procedure of extracting regions from an input image I_0 are as follows.

1. Set $k=1$.

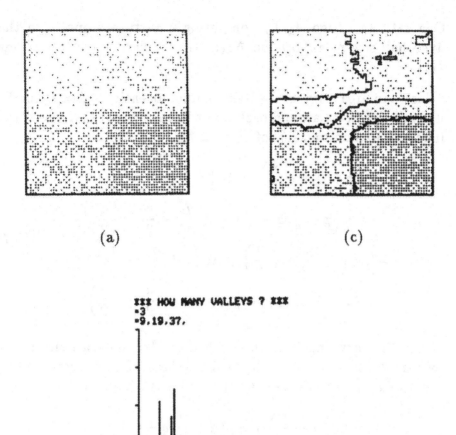

(a)

(c)

(b)

Figure 3.5: (a) Image containing four regions of different average intensity (probabilities of black points are 0.1, 0.2, 0.4, and 0.8, respectively); (b) Intensity histogram of image obtained by local averaging (a), using 16 × 16 neighborhood; (c) Regions obtained by thresholding locally averaged image (from Tomita and Tsuji, 1977).

2. Generate an image A_k by computing a texture property in the fixed half-size neighborhood $N_n(x, y)$ at each point in the image I_{k-1}.

3. Select one area having the least nonhomogeneity index from the five full-size areas around each point in the image I_{k-1}, where the index of nonhomogeneity of an area $N_{2n}(p, q)$ is

$$
\begin{aligned}
&\left| A_k \left(p + \frac{n}{2}, q + \frac{n}{2} \right) + A_k \left(p - \frac{n}{2}, q + \frac{n}{2} \right) \right. \\
&\left. - A_k \left(p + \frac{n}{2}, q - \frac{n}{2} \right) - A_k \left(p - \frac{n}{2}, q - \frac{n}{2} \right) \right| \\
&+ \left| A_k \left(p + \frac{n}{2}, q + \frac{n}{2} \right) + A_k \left(p + \frac{n}{2}, q - \frac{n}{2} \right) \right. \\
&\left. - A_k \left(p - \frac{n}{2}, q + \frac{n}{2} \right) - A_k \left(p - \frac{n}{2}, q - \frac{n}{2} \right) \right|
\end{aligned}
\tag{3.7}
$$

4. Generate a new image I_k in which the value of each point is the average texture value of the selected area of the corresponding point in I_{k-1}. The average texture value is computed by

$$
\begin{aligned}
&\left\{ A_k \left(p + \frac{n}{2}, q + \frac{n}{2} \right) + A_k \left(p - \frac{n}{2}, q + \frac{n}{2} \right) \right. \\
&\left. + A_k \left(p + \frac{n}{2}, q - \frac{n}{2} \right) + A_k \left(p - \frac{n}{2}, q - \frac{n}{2} \right) \right\} \bigg/ 4
\end{aligned}
\tag{3.8}
$$

5. Test whether all peaks in the histogram of I_k are sharp. If so go to step 7, or else set $k = k + 1$.

6. Generate an averaged image A_k by the fixed half-size neighborhood $N_n(x, y)$ of each point in the image I_{k-1}, and go to step 3.

7. Threshold I_k by the values of the valleys between the peaks in the histogram in order to extract regions.

The region in the selected area of each point is sometimes not the same as that of the point, however. The iterative smoothing operations correct this error and make the boundaries between the extracted regions smooth. Figure 3.7 illustrates the results of applying this selective

48

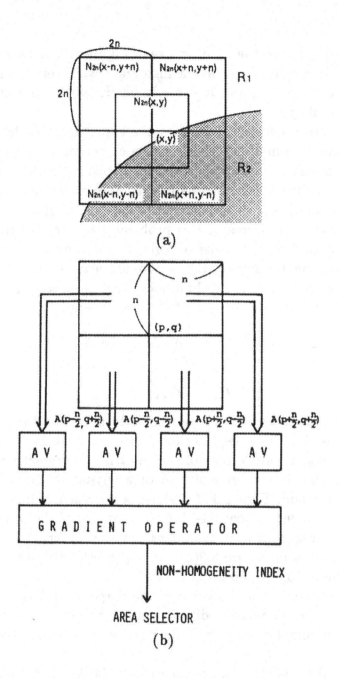

Figure 3.6: (a)Five areas around point (x, y) (at least one of them is entirely included in one region); (b) Calculation of nonhomogeneity index of $2n \times 2n$ area by applying gradient operator to averaged intensity levels in $n \times n$ subareas (from Tomita and Tsuji, 1977).

neighborhood method to the input image of Figure 3.5. After averaging three times, the output image has a histogram with very sharp peaks as shown in Figure 3.7(c), and it is easily partitioned into four regions with reasonable shapes.

Since the above algorithm assumes the smoothness of boundaries between regions, it sometimes gives erroneous results for regions with complex boundaries. For example, a wedge-shaped region with an acute angle in Figure 3.8(a) is likely to be merged into a surrounding region if its size is less than that of the neighborhood size, as shown in Figure 3.8(b). One method to overcome this problem is to vary the size of the neighborhood according to the size of regions at each point of the image.

At each point in the image, the nonhomogeneity index G_k is computed by changing the size of neighborhood $2k \times 2k$ $(k = 0, \ldots, L)$. The best value of k is determined by the following criterion.

$$G_L > \lambda G_{L-1} > \cdots > \lambda^{L-k} G_k \quad (\lambda = 0.75) \tag{3.9}$$

and

$$G_k \leq \lambda G_{k-1} \tag{3.10}$$

Figure 3.8(c) shows the result of segmentation using the neighborhood of variable size. The neighborhood sizes used are 17×17, 9×9, and 5×5. More acute angle portions of the triangle are extracted.

Figure 3.9 shows another example of applying the variable-sized neighborhood method. Figure 3.9(a) shows a microscopic image of eyeball cells. Figure 3.9(b) shows the intensity profile of the image along a line. It is averaged so that the edges between different regions are preserved, as shown in Figure 3.9(c). Then, thresholding the averaged image gives the regions in Figure 3.9(d).

The alternative solution is to change the shape of the neighborhoods from square to, for example, wedge. Nagao et al. (1979) proposed the edge preserving smoothing operator using this type of selective neighborhoods.

In spite of these efforts, the resulting boundaries of texture regions are still blurred because of the large neighborhoods. The last strategy is to combine the intensity-based segmentation and the texture-based segmentation. Consider the image in Figure 3.10(a). The intensity-based image segmentation gives the regions in Figure 3.10(b). There

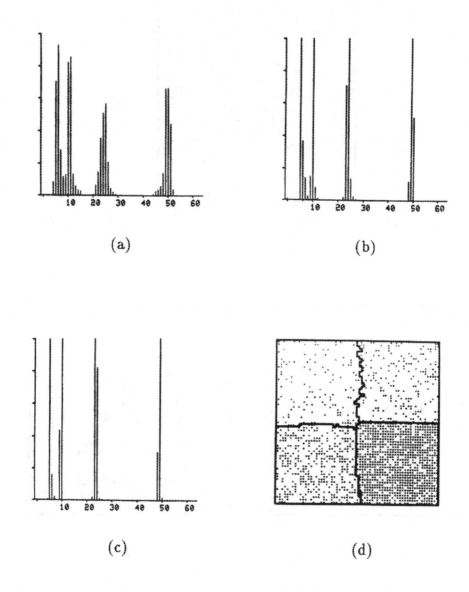

(a)

(b)

(c)

(d)

Figure 3.7: Results of applying the fixed-size nighborhood method to Figure 3.3(a): (a), (b), and (c) are intensity histograms of averaged images I_1, I_2, and I_3, respectively; (d) Regions obtained by thresholding I_3 (from Tomita and Tsuji, 1977).

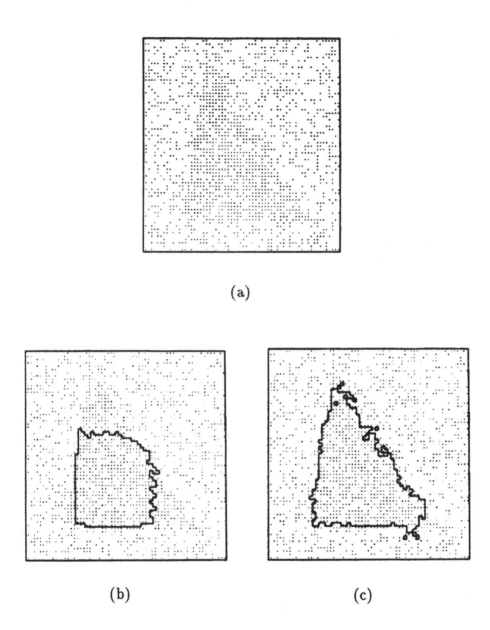

(a)

(b)   (c)

Figure 3.8: (a) Image containing two regions (triangle and background) of different average intensity (probabilities of black points are 0.7 and 0.3, respectively); (b) Result of applying the fixed-size neighborhood method; (c) Result of applying the variable-size neighborhood method.

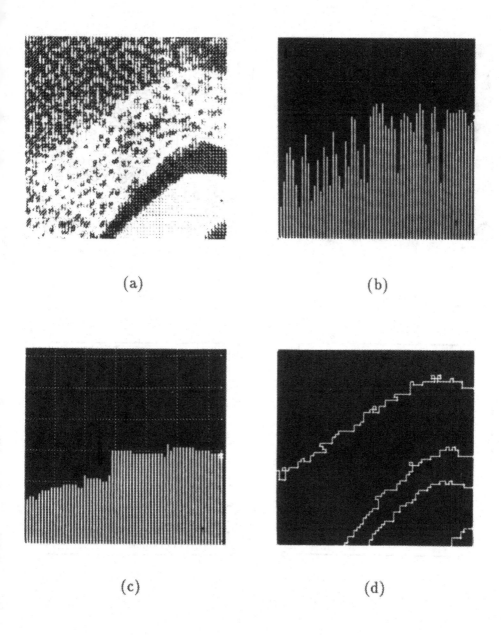

Figure 3.9: (a) Input image; (b) Original intensity profile; (c) Intensity profile after averaging; (d) Contours of regions.

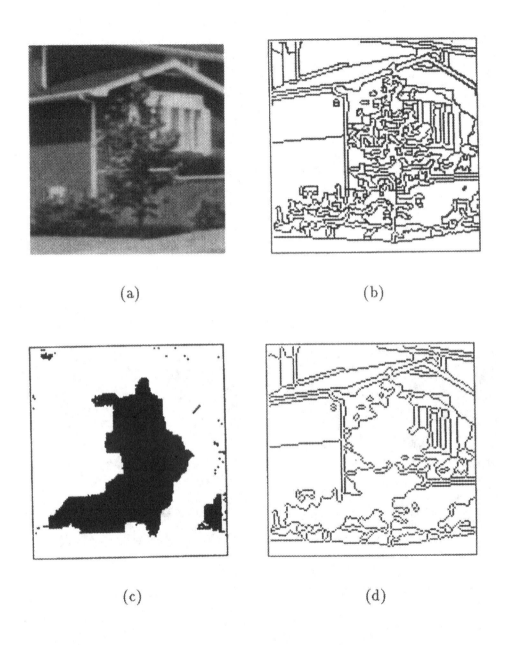

Figure 3.10: (a) Input image; (b) Regions with uniform intensity; (c) Regions with high edge-density; (d) Refined texture regions.

are many fine but tiny regions which correspond to a tree and bushes in the scene. The texture-based image segmentation, on the other hand, gives the regions with high edge-density, as shown in Figure 3.10(c). The region is large but blurred. Each fault is canceled out by overlapping both regions as follows.

1. Label the uniform regions with texture mark of which the area is more than fifty percent covered by the texture regions.

2. Merge the adjacent regions with the texture mark.

As a result, the deblurred texture regions are extracted, as shown in Figure 3.10(d).

3.3 Conclusion

A region has been defined as a connected component with similar intensity or texture. However, the problem of region segmentation requires more than simple analyzing image property values, because the goal of image segmentation is to find a segmentation that separates out meaningful objects or parts of objects. It is sometimes necessary, to further segment, merge, or group these regions geometrically or semantically to extract complex objects in complex images.

Chapter 4

Shape Analysis

Shape of a region extracted in the image is the important cue for recognizing the corresponding object as well as texture. Algorithms for shape analysis are classified under two criteria: whether they examine the boundary only or the whole area of the region, and whether they describe it in terms of scalar measurements or through structural description (Pavlidis, 1978). Accordingly, there are four basic algorithms for shape analysis: the scalar transform of the area which is useful for rough classification of shape, the Fourier transform of the boundary which is good at analyzing curved shape, the medial-axis transformation of the area which is good at analyzing elongate shape, and the segmentation of the boundary which is good at analyzing polygonal shape. This chapter presents the four basic algorithms for analysis of two-dimensional shape.

4.1 Scalar Transform

The shape of a region is described roughly in terms of the following scalar measurements, which are easy to compute and are appropriate as input to classical pattern recognizers.

1. **Area:** The area (A) of a region is defined by the number of points in the region.

2. **Perimeter:** The perimeter (P) of a region is defined by the number of points on the boundary of the region.

3. **Circularity:** The circularity (C) of a region is defined by $4\pi A/P^2$. Circular regions takes the maximum value 1.

4. **Size:** The size (S) of a region is defined by $2A/P$. The size of a circular region with radius r is r. The value of size is not affected greatly by the overlap of objects which make one region. For example, consider two circular objects which are overlapping and forming one region, as shown in Figure 4.1(a). The change of the area (ΔA) and that of the perimeter (ΔP) are represented by a function of angle θ, which is a degree of the overlap:

$$\begin{aligned} \Delta A(\theta) &= r^2(\theta - \sin\theta) \\ \Delta P(\theta) &= 2r\theta \end{aligned} \qquad (4.1)$$

Then, the size of the region is represented by

$$S(\theta) = \frac{2(A - \Delta A(\theta))}{P - \Delta P(\theta)} = r\left(1 - \frac{\sin\theta}{2\pi - \theta}\right) \qquad (4.2)$$

Figure 4.1(b) shows the curve of $S(\theta)$. We can see that the effect of overlap on the size is very small.

5. **Moments:** The moments around the center of gravity $(\bar{x}, \bar{y})$ of region R are defined by

$$M_{ij} = \sum_x \sum_y_{(x,y)\in R} (x - \bar{x})^i (y - \bar{y})^j \qquad (4.3)$$

Sample values of a set of moments $\{M_{20}, M_{02}, M_{11}, M_{21}, M_{12}\}$ are shown in Figure 4.2. These moments are sensitive to rotations. Rotation-invariant moments are computed by taking the region's major axis as the x coordinate axis.

6. **Principal Axes:** Let λ_1 and λ_2 $(\lambda_1 \geq \lambda_2)$ be the first and second eigen-values of the variance-covariance matrix of region R.

$$V = \begin{pmatrix} M_{20} & M_{11} \\ M_{11} & M_{02} \end{pmatrix} \qquad (4.4)$$

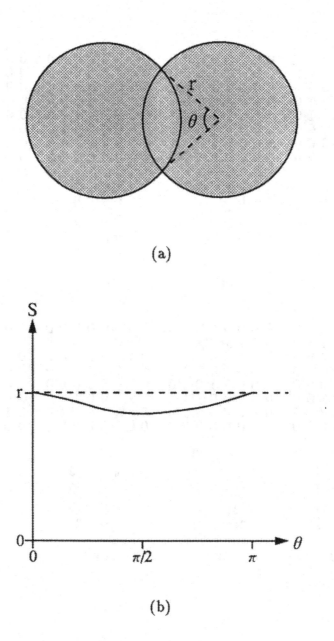

(a)

(b)

Figure 4.1: (a) Overlap degree of two circular objects; (b) Relation between size measure and degree of overlap.

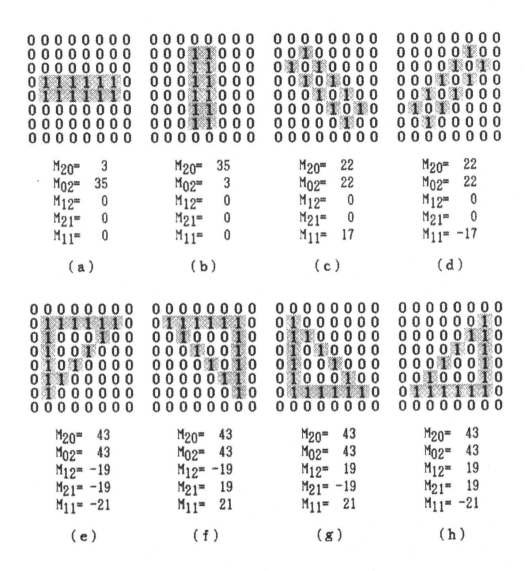

Figure 4.2: Moments.

Since λ_1 and λ_2 are the deviations of the points in the region from the two principal axes of the region, the ratio of the two eigenvalues defines the magnitude of *elongation* of the region:

$$L = \frac{\lambda_2}{\lambda_1} \tag{4.5}$$

And, the angular direction of the major axis defines the direction of elongation of the region:

$$\alpha = \frac{1}{2} \tan^{-1} \frac{2M_{11}}{M_{20} - M_{02}} \tag{4.6}$$

4.2 Fourier Transform

Zahn and Roskies (1972) proposed Fourier descriptors for the closed boundary of a region. Consider a boundary with parametric representation $P(l) = (x(l), y(l))$, where l is the arc length from the initial point on the boundary and $0 \le l \le L$. Let $\theta(l)$ be the angular direction of the tangent to the boundary at point $P(l)$. We define a function $\phi(l)$ as the rotation of the tangent at the point on the boundary from $\theta(0)$ to $\theta(l)$, as shown in Figure 4.3. When the points on the boundary are ordered clockwise, $\phi(L) = -2\pi$. Then, the function can be normalized to a periodic function $\phi^*(t)$ by

$$\phi^*(t) = \phi \left(\frac{Lt}{2\pi} \right) + t \tag{4.7}$$

where the domain of the function is $[0, 2\pi]$ and $\phi^*(0) = \phi^*(2\pi) = 0$.

The ϕ^* is invariant under translations, rotations, and changes of perimeter L. We now expand ϕ^* as a Fourier series. In polar form the expansion is

$$\phi^*(t) = \mu_0 + \sum_{k=1}^{\infty} A_k \cos(kt - \alpha_k) \tag{4.8}$$

These numbers A_k and α_k are the Fourier descriptors for the boundary and are known respectively as the kth harmonic amplitude and phase angle. The lower-order terms reflect the global shape of the bounadary and the higher-order terms are affected by small fluctuations of the boundary.

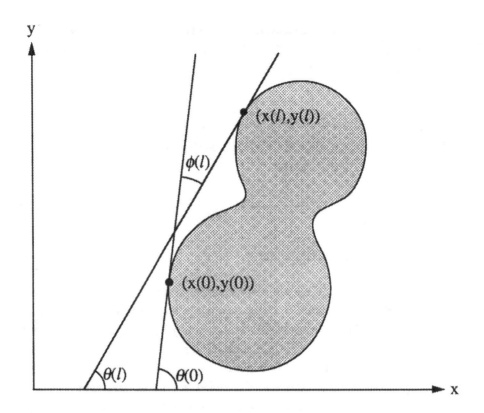

Figure 4.3: Parametric representation of a curve with tangential direction $\theta(l)$ and cumulative angular bend function $\phi(l)$.

4.3   Medial-Axis Transformation

The shape of a region is represented by the medial-axis of the region and the cross sections along the medial-axis. The coross section is perpendicular to the medial-axis and is approximately the distance to the nearest point on the boundary of the region. The medial-axis of a region, which was proposed by Blum (1964), is extracted by *region thinning*, and the distances from the points on the medial-axis to the boundary are obtained by *distance transformation*. A complex medial-axis is segmented into simple sub-medial-axes. Since the connectivity of the region is preserved in the medial-axis, segmentation of the medial-axis corresponds to *decomposition* of the region.

4.3.1 Distance Transformation

Distance transformation is an operation which compute the distance of each point in the region from the nearest point in the background. The following 4-neighbor distance d_4 or 8-neighbor distance d_8 between two points in the image is used instead of general Euclidean distance for easy computation (Rosenfeld and Pfalts, 1968).

$$d_4((i,j),(h,k)) = |i - h| + |j - k|$$
$$d_8((i,j),(h,k)) = \max\{|i - h|, |j - k|\} \tag{4.9}$$

A sequential algorithm for computing 4-neighbor distance $D(i,j)$ of each point (i,j) in region R from the nearest point in the background is as follows.

1) Initialize f and u as follows:

$$\begin{aligned} \text{if } (i,j) \in R, \quad f(i,j) &= 1 \quad \text{and} \quad u(i,j) = \infty \\ \text{if } (i,j) \notin R, \quad f(i,j) &= 0 \quad \text{and} \quad u(i,j) = 0 \end{aligned} \tag{4.10}$$

2) Scanning f in the forward direction in Figure 4.4(a), compute u according to the following rule:

$$u(i,j) = \min\{f(i,j), u(i-1,j)+1, u(i,j-1)+1\} \tag{4.11}$$

3) Scanning u in the backward direction in Figure 4.4(b), compute D according to the following rule:

$$D(i,j) = \min\{u(i,j), D(i-1,j)+1, D(i,j-1)+1\} \tag{4.12}$$

For example, Figure 4.5 shows the result of the distance transformation of a given region.

4.3.2 Region Thinning

The medial-axis of a region, which is extracted by repeatedly thinning the region, must satisfy the following requirements (Hilditch, 1969):

- Thinness: the medial-axis is one-pixel wide,

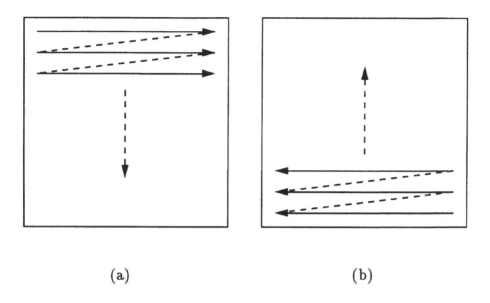

Figure 4.4: (a) Forward raster scan; (b) Backward raster scan.

Figure 4.5: Distance transformation.

r_6	r_7	r_8
r_5	r_0	r_1
r_4	r_3	r_2

Figure 4.6: The 8-neighbors of point r_0.

- Position: the medial-axis lies along the centers of the region,

- Connectivity: the medial-axis preserves the connectivity of the region, and

- Stability: as soon as part of the medial-axis is obtained, a point which lies at the tip of the axis must not eroded away by subsequent operation.

Many algorithms have been proposed for region thinning (Tamura, 1978). The basic operation is to remove any point of the region which does not destroy the connectivity of the region and which is not the terminal point of the axis. This operation is applied repeatedly from the boundary points to the interior points of the region until no point is removed.

Let r_k $(k = 1, ..., 8)$ denote the 8-neighbors of point r_0 in region R, as shown in Figure 4.6, and $f(r_i)$ be defined as

$$
\begin{aligned}
\text{if } r_i \in R, \quad f(r_i) &= 1 \\
\text{if } r_i \notin R, \quad f(r_i) &= 0
\end{aligned}
\tag{4.13}
$$

The *connectivity number* $N_c^{(4)}$ or $N_c^{(8)}$ of point r_0 is useful to test if removing it destroys 4-connectivity or 8-connectivity of the current region R, respectively:

$$
\begin{aligned}
N_c^{(4)}(r_0) &= \sum_{k \in \{1,3,5,7\}} \{f(r_k) - f(r_k)f(r_{k+1})f(r_{k+2})\} \\
N_c^{(8)}(r_0) &= \sum_{k \in \{1,3,5,7\}} \{\overline{f}(r_k) - \overline{f}(r_k)\overline{f}(r_{k+1})\overline{f}(r_{k+2})\}
\end{aligned}
\tag{4.14}
$$

65

Figure 4.7: Region thinning.

where $\overline{f}(r_i) = 1 - f(r_i)$ and $r_9 = r_1$. The connectivity number ranges from 0 to 4. It is proved that the point which does not destroy the connectivity has the connectivity number 1 (Yokoi et al., 1981). For example, Figure 4.7 shows the result of thinning the region in Figure 4.5.

4.3.3 Region Decomposition

When the shape of a region is complex, the medial-axis also takes a complex form. A complex medial-axis is segmented into sub-medial-axes each of which is a simple line or curve segment and has a simple group of smoothly changing cross sections on it. The points of segmentation on the medial-axis are the branching points, the sharply bending points, and the transitional points between different groups of cross sections. These points are detected as follows.

1. The medial-axis is first segmented into sub-axes at the branching points, as shown in Figure 4.8(a). If the two sub-axes is smoothly connected at the point, however, they are merged into one axis, as

66

shown in Figure 4.8(b). The smoothness measure is the amount of change of the tangents at the two terminal points. If the change of the tangent is low, the pair of axes may be merged.

2. If there is a sharply bending point, i.e., a vertex on the axis, it is segmented at the vertex, as shown in Figure 4.8(c). The procedure to detect vertices of the axis is the same as the boundary segmentation in the following section.

3. There are two types of trasitional points between different cross sections. Let $\{d_0, \ldots, d_n\}$ denote the cross sections of one axis. One type of transitional point is detected by finding the local minimum of the cross sections along the axis, as shown Figure 4.8(d). If the axis has a point i which satisfies the following condition, the axis is segmented at the point.

$$d_i = \min\{d_{i-k}, \ldots, d_i, \ldots, d_{i+k}\}$$

$$\frac{d_{i-k} + d_{i+k} - 2d_i}{2kd_i} > t_1 \qquad (4.15)$$

4. The other type of transitional points are detected by finding monotonically decreasing cross sections from a wide group to a narrow group, as shown in Figure 4.8(e). This type of points are also detected on the axis called *spur* which does not contribute to the principal shape of the region. Spurs are caused by small fluctuations of the boundary. They should be removed as noise. If the axis has a point which satisfies the following condition, the point is removed, and accordingly the axis is segmented there.

$$d_{i-k} \geq \cdots \geq d_i \geq \cdots \geq d_{i+k}$$

$$\frac{d_{i-k} - d_{i+k}}{2k} > t_2 \qquad (4.16)$$

As a result, the region is represented by a graph; the node denotes a sub-axis and the link denotes the original connectivity between two sub-axes. Each sub-axis can be used to reconstruct the corresponding region which is a subset of the original region by expanding each sub-axis to the

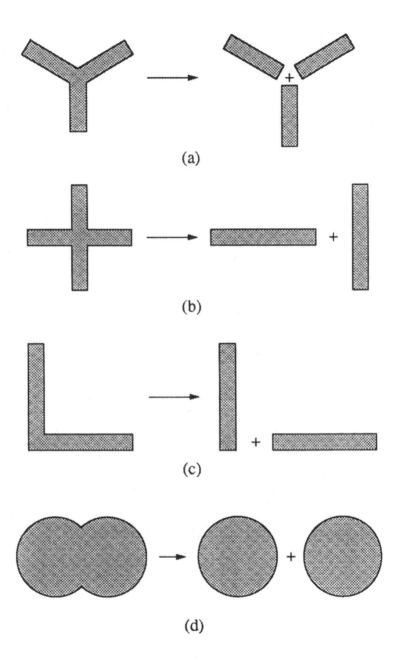

Figure 4.8: Region decomposition.

original width independently. *Region expansion* is the reverse operation of region thinning. The examples are shown in the next chapter.

4.4 Boundary Segmentation

The most common algorithm for shape analysis is to segment the boundary of a region into the simple line or curve segments. The typical points of segmentation of the boundary are vertices of the boundary, the sharply bending points. The vertices of the boundary are detected by computing the curvature at each point on the boundary. The *k-curvature* at point Q in Figure 4.9 is defined by the difference in angle between PQ and QR, where P and R are *k* points away from Q. If the curvature is locally maximal or minimal and the absolute value of the curvature is large, it is regarded as the vertex of the boundary.

This method only, however, sometimes fails to detect proper vertices because of local fluctuations of the boundary. The method to overcome this problem is to fit a straight line or a simple curve (for example, ellipse) to each segment by a least square method. If the fitting error is large, the segment is further split into (sub)segments at the point which

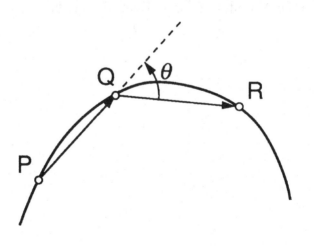

Figure 4.9: Curvature.

is the farthest from the line defined by the two terminal points of the segment. Another line or curve is again fitted to each of the two new segments. This process is repeated until the fitting error becomes low, which is called an *iterative endpoint fit* (Duda and Hart, 1973).

As a result, the boundary is represented by a list of ordered segments each of which is a simple line or curve segment.

4.5   Conclusions

We have overviewed four common algorithms for shape analysis. Since each algorithm has advantage and disadvantage, the proper algorithm must be used depending on the purpose of the visual task. The simplest scalar transforms of regions play a role of pure perception or filtering of shape. The Fourier transform is powerful for classification of given types of shape. The disadvantage of all transform techniques is the difficulty of describing local information. The medial-axis transformation is especially useful for describing elongate shape and the representation is sometimes called *ribbon* as the projection of generalized cylinder, one solid model of three-dimensional shape. The boundary segmentation is especially useful for describing polygons and produces the two-dimensional version of *B-reps* (boundary representations), the most common solid model of three-dimensional shape.

Chapter 5

Structural Texture Analysis

From the structural view of texture, texture is considered to be composed of texture elements; texture is defined by the texture elements which are arranged according to a placement rule. The analysis includes extraction of texture elements in the image, shape analysis of texture elements, and estimation of the placement rule of texture elements. Thus the structural texture analysis is complex in camparison with the statistical texture analysis, and it gives more detailed description of texture. This chapter presents the algorithms for structural texture analysis.

5.1 Texture Elements

The first step of structural texture analysis is to extract texture elements in a given texture image. A texture element is defined by a uniform intensity region of simple shape in the image. The procedure of extracting texture elements in the image is as follows.

1. Segment the image into uniform intensity regions. For example, the images in Figure 5.1 are segmented into uniform regions by the merging-and-thresholding method, as shown in Figure 5.2.

2. Extract the medial-axes of regions by region thinning and distance transformation, as shown in Figure 5.3.

71

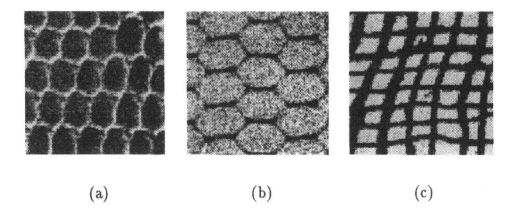

(a)   (b)   (c)

Figure 5.1: Texture images: (a) Reptile skin; (b) Netting; (c) Loose burlap.

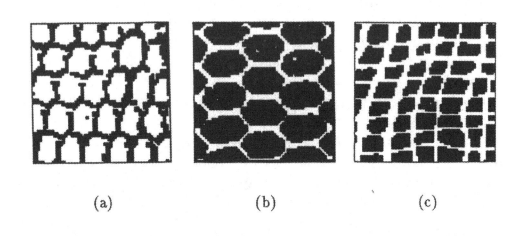

(a)   (b)   (c)

Figure 5.2: Uniform regions extracted in the images.

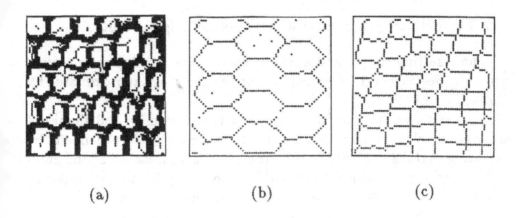

(a)   (b)   (c)

Figure 5.3: Medial-axes of regions.

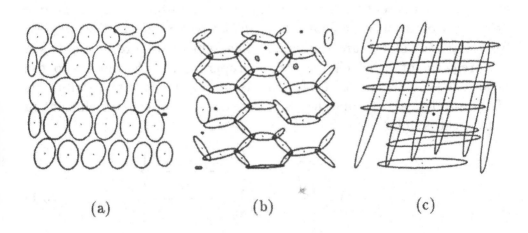

(a)   (b)   (c)

Figure 5.4: Atomic texture elements (regions are approximated by ellipses only for display).

3. Segment the complex axes into the simple subaxes at the branching points, the bending points, or the transitional points, and label the resulting subaxes with different numbers.

4. Expand each subaxis to the region of the original width.

The resulting simple and uniform regions are called *atomic texture elements*. See Figure 5.4 in which texture elements are approximated by ellipses.

The next step is to compute the properties (intensity and shape) of each texture element and to find the sets of similar texture elements by clustering texture elements in the property space. Detailed shape analysis of each texture element is usually unnecessary for texture discrimination. For example, Figure 5.5 illustrates the sets of similar texture elements in Figure 5.4 which are based on such simple scalar measurements as area, size, elongation, and direction of texture elements. Let (μ_{ij}, σ_{ij}) be the mean and the deviation of the jth property of texture elements in the ith set. Such a distribution is used for texture discrimination.

When the detailed shape analysis of the texture elements is needed, for example, different textures have the same distributions of the simple properties of texture elements, one *typical texture element* is selected from each set of texture elements. The typical texture element is a texture element of which the shape properties are the closest to the mean values in the set it belongs to. Assuming the normal distributions of the properties, the function to select the typical texture element in the ith set is defined by

$$C_i(\{x_j\}) = \frac{\sum_{j=1}^{n} a_j \exp\left\{-\frac{(x_j - \mu_{ij})^2}{2\sigma_{ij}^2}\right\}}{\sum_{j=1}^{n} a_j} \tag{5.1}$$

where x_j is the value of the jth property of a texture element, a_j is the weight on the jth property. And, the detailed shape analysis (namely, medial-axis transform or boundary segmentation) is performed to the typical texture element which has the minimum value of C_i.

74

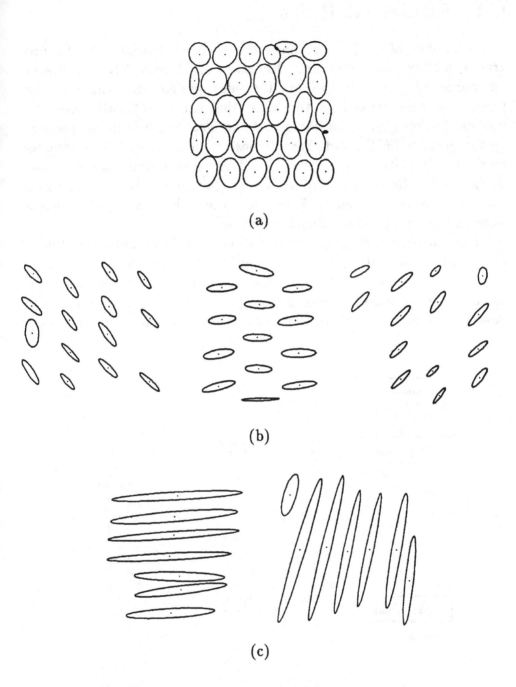

(a)

(b)

(c)

Figure 5.5: Sets of similar texture elements.

5.2 Placement Rules

A number of rules to arrange the texture elements can be programmed to synthesize textures (Rosenfeld and Lipkin, 1970). However, our purpose is inversely to know the rules by analyzing unknown textures. The rules must have the forms that can be estimated from the images. Zucker (1976) considers that the *ideal textures* are represented by the graph (GRIT), and that the *observable textures* are represented by the graph (GROT) which is modified by a transformation τ, as shown in Figure 5.6. He proposed to use the graph which is isomorphic to the three regular tesselations in Figure 5.7, the eight semi-regular tesselations in Figure 5.8, or the linear subgraph.

The most common regular tesselations and linear subgraph can be estimated by computing the relative positions between texture elements in the image. More complex rules are analyzed hierarchically later. Let us compute the two-dimensional histogram of the relative positions between every pair of texture elements. If the placement of the texture

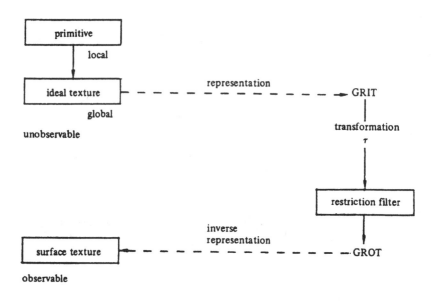

Figure 5.6: Ideal texture and observable texture (Zucker, 1976).

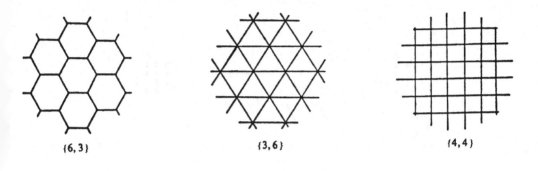

Figure 5.7: Regular tesselations (Zucker, 1976).

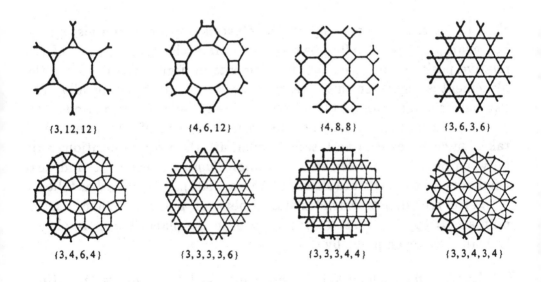

Figure 5.8: Semiregular tesselations (Zucker, 1976).

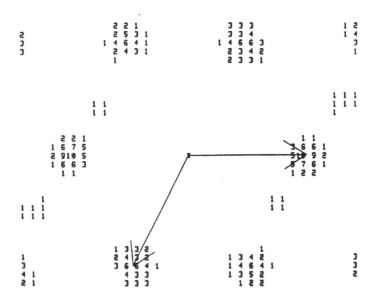

Figure 5.9: Two-dimensional histogram of relative position vectors between texture elements in Figure 5.5(a).

elements is regular, some significant clusters appear in the histogram. For example, Figure 5.9 shows the two-dimensional histogram of the relative positions between texture elements in Figure 5.5(a). Since the placement is regular in this case, some clusters appear periodically. The two base clusters around $(11, 0)$ and $(-6, -13)$ specify the periods.

Computing relative positions between every pair of texture elements takes much time. It is necessary to compute the relative positions only between near texture elements. *Voronoi diagram* is useful to find near neighbors for each texture element (Ahuja, 1982). It is defined as follows.

Let $d(p, q)$ denote Euclid distance between p and q. When a set of points $P = \{p_1, p_2, \ldots, p_n \in \mathbf{R}^2\}$ is given in the plane, Voronoi polygon is defined for each point by

$$V(p_i) = \{p \in \mathbf{R}^2 | d(p, p_1) < d(p, p_j) \text{ for } j = 1, 2, \ldots, n; j \neq i\}. \quad (5.2)$$

The set of Voronoi polygons $\{V(p_1), V(p_2), \ldots, V(p_n)\}$ is called Voronoi diagram. For example, the solid lines in Figure 5.10(a) show Voronoi diagram defined by a given set of points.

78

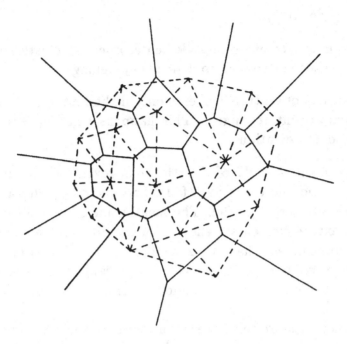

Figure 5.10: Voronoi diagram (solid lines) defined by a given set of points and the corresponding Delaunay triangulation (dotted lines) (from Ahuja, 1982).

If two Voronoi polygons have a common edge, the corresponding points are considered as the neighbors of each other. Connecting every pair of neighboring points produces *Delaunay tiangulation*, which is a dual graph of Voronoi diagram. The dotted lines in Figure 5.10(b) show Delaunay triangulation.

Then, the procedure to detect the base clusters is as follows.

1. Compute the first relative position vector $v_1(r_1, \theta_1 \bmod \pi)$ of each texture element to the nearest texture element in the same set.

2. Detect clusters in the histograms of the first vectors r_1 and θ_1.

3. Compute the second relative position vector $v_2(r_2, \theta_2 \bmod \pi)$ of each texture element to the independently second nearest texture element which is not included in the clusters of the first vectors.

4. Detect clusters in the histograms of the second vectors r_2 and

$\theta_2 - \theta_1 \pmod{\pi}$.

5. Make new sets of texture elements based on clusters of the first and the second vectors to which they belong.

6. Merge sets of texture elements only if the first vectors and second vectors are interchanged. This happens when $r_1 \cong r_2$ and they have some deviations.

The placement rule Φ_i of the ith set of texture elements is defined by the mean and the deviation of the first vector (r_1, θ_1) and those of the second vector $(r_2, \theta_2 - \theta_1)$. The mean value of r_1 is an index of the *density* of texture elements; dense textures will have short vectors and sparse texture elements have long vectors. The deviation of each vector is an index of *regularity* of the placement; regularly arranged textures will have a small deviation and randomly arranged textures have a large deviation.

When there is more than one set of texture elements, the relative position vector to the nearest texture element is computed for each texure element between each pair of sets of texture elements. Then, the smaller sets of texture elements may be further created using these vectors. The placement rule Φ_{ik} which specifies the relative position between a pair of texture elements, one in i set and the other in the kth set, is defined by the mean and the deviation of these vectors.

5.3 Hierarchical Textures

There are hierarchical textures in which a set of texture elements form a bigger unit (subpattern) and such units are arranged according to another placement rule. See, for example, two textures in Figure 5.11. Although both textures have the same kinds of texture elements and the same relative positions, four neighboring texture elements in Figure 5.11(b) form a larger unit. These large units must be detected to discriminate these two textures, as shown in Figure 5.12.

The process to detect such units is called *grouping*, which will be discussed in Chapter 6. Since a set of units may form a further larger unit, the grouping process is repeated until no larger unit is formed.

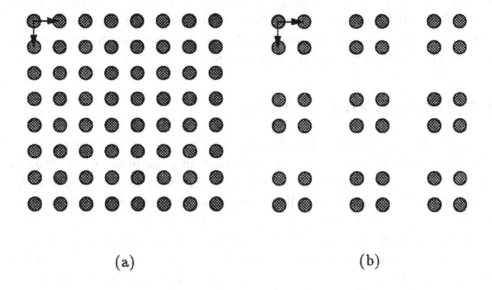

(a)   (b)

Figure 5.11: (a) Plain texture; (b) Hierarchical texture.

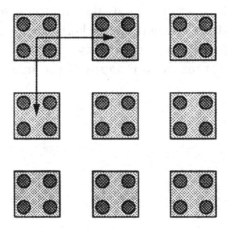

Figure 5.12: Grouping of texture elements.

As a result, the texture is described hierarchically. At the top level the texture is described by the largest units with their placement rules. Each unit is described by the units on a lower level with their placement rules, and so on. On the bottom level, each unit is described by the atomic texture elements and their placement rules.

5.4 Figure and Ground

Note that there are generally at least two sets of texture elements in the image; one is the *figure* of the image and the other is the *ground* of the image. Since the structure of the figure uniquely determines the structure of the ground, detailed structural description is needed only for the figure of the image. There is no matter if only one set of texture elements is found in the image. If more than one set of texture elements are found in the image, on the other hand, one of them can be regarded as the ground of the image. If the observer has knowledge about the object in the image, he will select the figure of the image properly. On the other hand, if he knows nothing about the object, there is no unique way to determine the figure or the ground. Psychologically, regular patterns are apt to be the figure of the image. It is possible to rank the regularity of each set of texture elements according to the deviations of the properties of texture elements and their placement rules. A set of texture elements with the lowest regularity will be selected as the ground of the image and the others as the figure of the image.

Chapter 6

Grouping

When there are more than one kind of texture in the image, statistical texture properties has been computed locally at each point in the image, and then image segmentation has been applied to extract homogeneous regions with consistent texture properties. When more than one kind of texture elements are detected in the image by the structural texture analysis, in contrast with image segmentation, *grouping* is applied to extract homogeneous regions each of which is covered with the same texture elements arranged by the same placement rule. Grouping is a process to find clusters of texture elements not only in the property domain but also in the spatial domain. This chapter presents the algorithms for grouping texture elements in the image.

6.1 Basic Operations

Let's consider a scene in Figure 6.1, which includes a cube and a table covered by different texture elements. Texture elements in the image are extracted by image segmentation and region decomposition, about which we have discussed in Chapter 5. Figure 6.2 shows texture elements extracted simply by the thresholding method. And, grouping will separate the texture elements to those on each surfaces of the cube and the table. The basic operations of grouping are as follows.

1. Compute the properties of texture elements, and then classify texture elements into sets of texture elements so that the texture

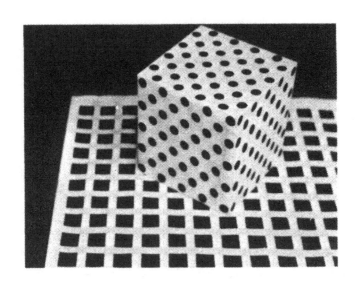

Figure 6.1: A cube on a table.

Figure 6.2: Texture elements detected in the image of Figure 6.1.

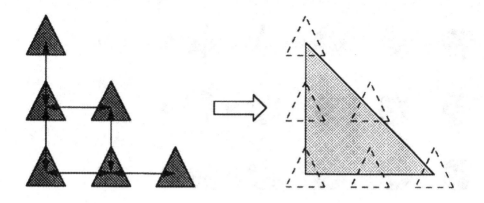

Figure 6.3: Area covered by texture elements of a region.

elements have the consistent properties.

2. Compute the relative positions v_1 and v_2 for each texture element to the nearest and the independently second nearest texture elements in the same set, respectively, and then further classify texture elements into subsets of texture elements so that the texture elements have the consistent relative poisitions, in another word, the texture elements are arranged according to the consistent placement rule (defined by the ranges of v_1 and v_2).

3. Test *connectivity* of texture elements in each set, respectively, since there may be texture elements that have the same placement rule but form spatially separate regions. Let us denote Φ_i the placement rule of the ith set of texture elements. Any pair of texture elements in the set are linked if the relative position is within the ranges of Φ_i. As a result, a set of linked texture elements form a region, as shown in Figure 6.3, and another set forms another separate region if exists.

4. Test *overlap* between regions. Figure 6.4 shows a simple example in which there exist two sets of texture elements. Let R_1 and R_2 be regions covered by small dots and large dots, respectively. Since

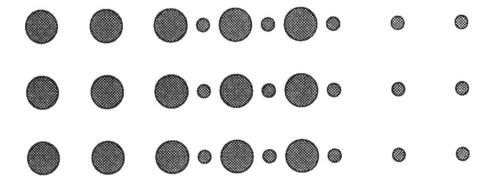

Figure 6.4: An example of three regions constructed by two sets S_1 and S_2.

the common area $R_1 \cup R_2$ of two regions has considerable area, it seems reasonable to regard these areas as three regions $R_1 \cap R_2$, $R_1 \cap \overline{R}_2$, and $\overline{R}_1 \cap R_2$, instead of two.

6.2 Recursive Grouping

A block diagram of the homogeneity analyzer for grouping is shown in Figure 6.5. First of all, the properties (descriptors) of texture elements are computed and the list is constructed. Second, histograms of these properties are computed. The supervisor examines these frequency distributions in order to select the most promising property for classifying the texture elements into groups in the property domain. We considered that a property is promising for initial partitioning if the texture elements are definitely classified into a few groups by it. Choosing the properties whose histograms have deep valleys between prominent peaks, the supervosor sets thresholds at the bottoms of the valleys. Since the validity of classifying the members in the valleys by thresholding is ambiguous, the supervisor selects the property having the minimum number of such members. Figure 6.6 shows the histograms of the prop-

86

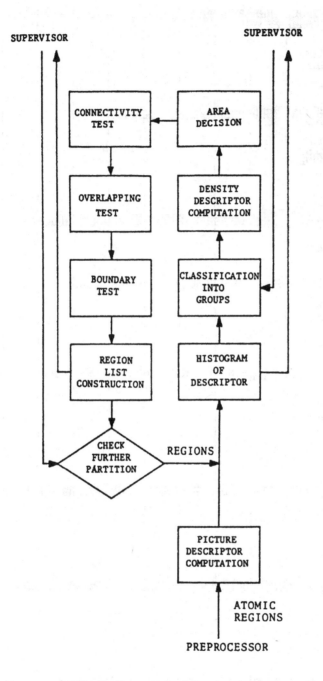

Figure 6.5: Block diagram of homogeneity analyzer.

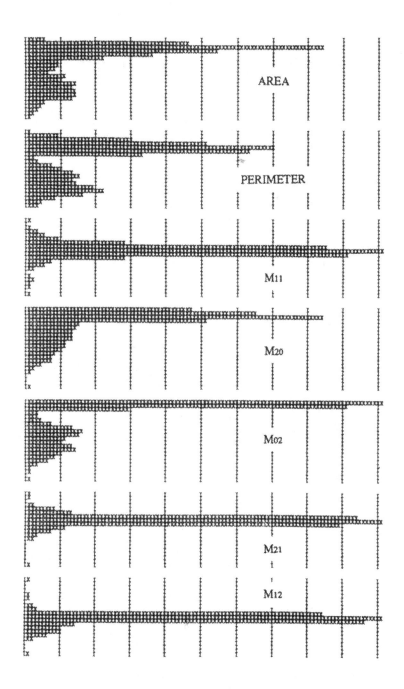

Figure 6.6: Histograms of geometrical properties of texture elements in Figure 6.2.

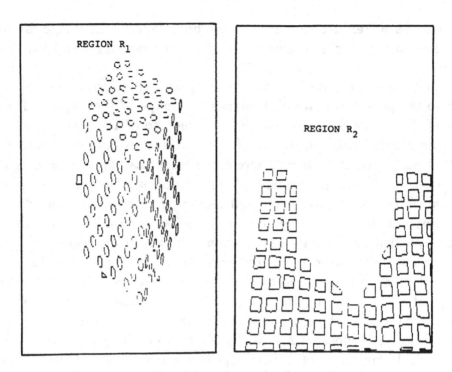

Figure 6.7: Partitioned results by perimeters.

erties of the texture elements in Figure 6.2. The perimeter is selected
for first-stage partitioning.

Next, the placements of the members of each group are evaluated. If
some members have relative positions whose values are isolated, they are
regarded as isolated texture elements and are excluded from the group.
As mentioned in the previous section, the areas covered by these groups
are determined, and then their connectivity and overlap are examined
to define regions. If a connected area of region candidates is small, the
analyzer does not regard it as a region and registers its members in a
group of isolated texture elements. Next, a boundary test is done to
check whether any texture elements in the group of isolated texture elements are located between two regions or surrounded by a region. They
are joined to the region if only one region touches them. However, if
two regions touch them, then they are left in the isolated group pend-
ing further verification. Figure 6.7 shows the results of grouping. The
regions R_1 and R_2 correspond to the cube and the table, respectively.

The small-size texture elements at the boundary of the image are not classified as region candidates of R_2, but the boundary test joins them to R_2.

The regions described by the boundaries and texture elements are sent to the supervisor, which analyzes the shape of the regions. If any part of the region does not match the structure of a model, the supervisor consideres that the result of the partition is not successful, and further partitioning is done. There are three strategies for changing the partition: (1) deeper partitioning, (2) modified partitioning after changing threshold values, (3) new partitioning using another property. In the experiments, only deeper partitioning has been done; thus, the iterative partitioning is terminated when any part of the regions is found to match a model structure or no further partitioning is possible.

Regions $R_1, R_2, \ldots$ found by the first stage partitioning are processed independently to obtain deeper partitions $R_{11}, R_{12}, \ldots, R_{21}, R_{22}, \ldots$. The procedures are the same as those in the first stage. Figure 6.8 shows the histogram of moment M_{11} in R_1 (Cube), in which three groups are observed. It must be noted that M_{11} of Figure 6.6 has no deep valley. This shows that the successive procedures can detect finer differences between members having an almost identical property, if the previous partition is adequate. Figure 6.9 shows the results of partitioning R_1 into three regions R_{11}, R_{12}, R_{13} which correspond to three faces of the cube. Two small triangles and a square at the boundary of R_1 in Figure 6.2 are classified into the isolated texture element group and eliminated

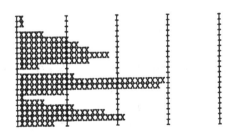

Figure 6.8: Histogram of M_{11} in R_1 (cube).

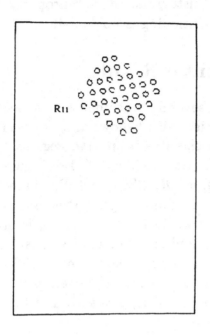

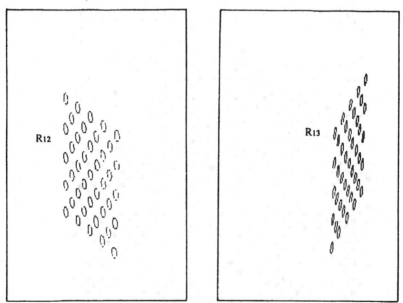

Figure 6.9: Further partitioning R_1 into three regions.

from the regions. The histograms of each property in R_2 have no deep valleys, and further partitioning is not possible.

6.3   Experimental Results

Experiments using several input images have been done to check the validity of grouping. Figure 6.10 shows an example in which six regions exist. The system first classifies it into two regions R_1 and R_2 using the area descriptor. The results are illustrated in Figure 6.11. R_1 is further divided into three regions R_{11}, R_{12}, and R_{13} based on the placement rules. R_{12} and R_{13} have the same properties but are separated by the connectivity test, as shown in Figure 6.12. R_2 is divided into R_{21} and R_{22} by examining M_{12}, and then the relative positions between texture elements are checked for further partitioning. There are four regions $R_{211}, R_{212}, R_{221}$, and R_{222}; however, the overlap test joints regions R_{212} and R_{222} and obtains a region R_{23}, as shown in Figure 6.13.

Figure 6.10: An input image having six regions.

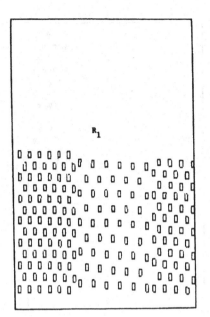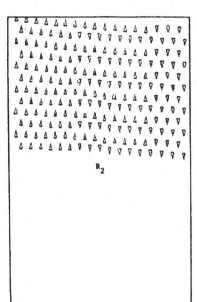

Figure 6.11: First-stage partitioning into R_1 and R_2.

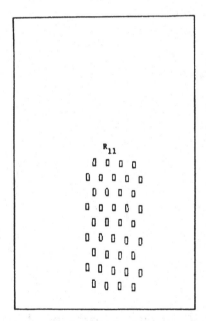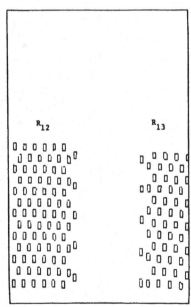

Figure 6.12: Second-stage partitioning R_1 into R_{11}, R_{12}, and R_{13}.

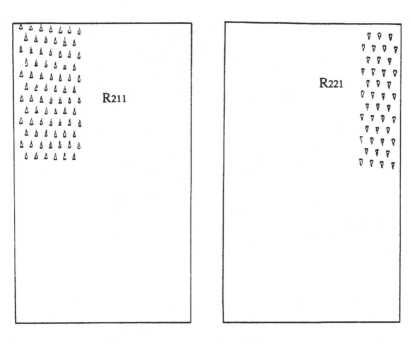

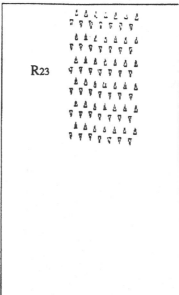

Figure 6.13: Second- and third-stage analysis yeild a partition of R_2 into three regions R_{211}, R_{221}, and R_{23}.

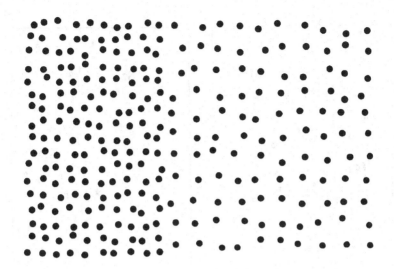

Figure 6.14: An input image in which texture elements are randomly arranged.

Another example of an input image is shown in Figure 6.14. The texture elements in this example are randomly arranged. Some of the histograms of the properties of texture elements are shown in Figure 6.15. Testing these data, the density property (length of the first relative position vector v_1) is applicable to partitioning. Figure 6.16 shows the results of grouping the texture elements into two sets S_1 and S_2 in which several isolated texture elements exist. Using the boundary test,

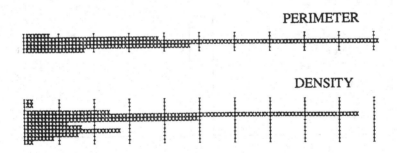

Figure 6.15: Some of histograms of the properties of texture elements.

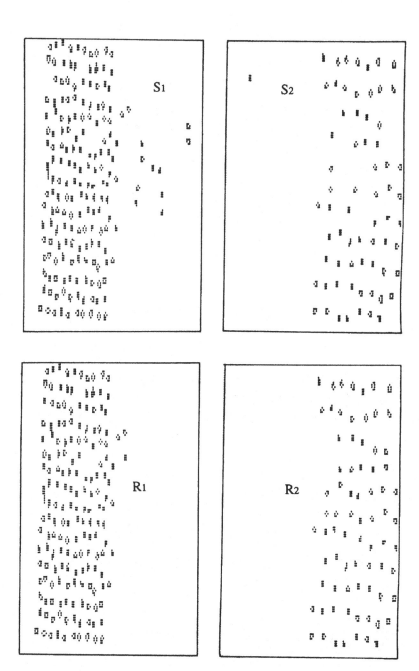

Figure 6.16: Partition of the image in which texture elements are arranged randomly.

the system merges them into regions and obtains two regions R_1 and R_2 shown in Figure 6.16.

Chapter 7

System and Evaluation

There are two approaches to texture analysis: statistical analysis and structural analysis. The difference is the level of description of textures. Statistical texture analysis gives simple scalar measurements of textures. Structural texture analysis, on the other hand, gives detailed structural descriptions of textures. Which level of analysis should be used? Comparatively, statistical analysis is suitable for fine microtextures and structural analysis for coarse macrotextures. Either approach alone cannot deal with all kinds of textures properly. In the first section of this chapter, *adaptive texture analysis* is presented as a total system for texture analysis, which uses either or both of statistical and structural analyses depending on nature of texture, to analyze variety of textures.

One purpose of texture analysis is to classify images of textures into several categories. For example, there is a need to classify images of aerial photographs according to the terrains. Then, a conventional way for testing performance of texture analyzers is to evaluate their classification abilities using texture samples of which the categories are specified by the user. The best analyzer is that which achieves the highest score in classification (Weszka et al. 1976a; 1976b; Conners and Harlow, 1980). However, classification ability is highly dependent on both the types of textures and the specified categories. Since not only each analyzer has merits and demerits but also the categories of natural textures are not as definite as those of artificially made characters, the resulting evaluation

99

of analyzers changes according to the field of application.

Anyway, there are two phases in classifying texture images: the learning phase to give the standard description of each category of texture and then the recognition phase to match the description of any input image with the standard description of each category. In the second section of this chapter, the procedure of each phase of image classification is presented when analyzing textures structurally.

One way to evaluate the absolute ability of a texture analyzer is *analysis-by-synthesis* (Tomita, 1982). It enables to see the total information included in the output description in terms of image. The corresponding synthesizer or the inverse transformer generates a texture image from the description. By comparing the reconstructed image with the original image, we can see what information is preserved and what is lost in the description. Generally, the more information there is in the description, the better is the reconstructed result. For example, the original image can be reconstructed if all components of Fourier transform are known. However, they are not all useful for image classification. There is a trade off for texture analyzers between classification and synthesis.

In the last section of this chapter, the procedure to reconstruct textures using the descriptions given by the structural analysis is presented, and some other *generative models* of textures are reviewed briefly.

7.1 Adaptive Texture Analysis

There is no unique way to analyze every texture. Even when analyzing the same texture, the procedure varies according to the purpose of the analysis, the types of textures to be analyzed together, and the categories of textures specified by the user. It is desirable for the texture analysis system to be able to deal with any situation. The adaptive texture analysis has been proposed to analyze textures statistically or structurally by recognizing the types of textures in the course of analysis (Tomita, 1981).

Consider four illustrative textures in Figure 7.1 and the corresponding real textures in Figure 7.2. The typical way to analyze each of these patterns is as follows.

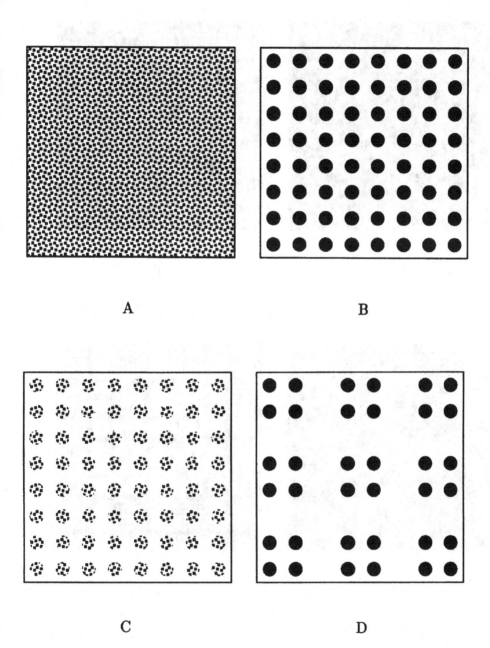

A

B

C

D

Figure 7.1: Four illustrative textures.

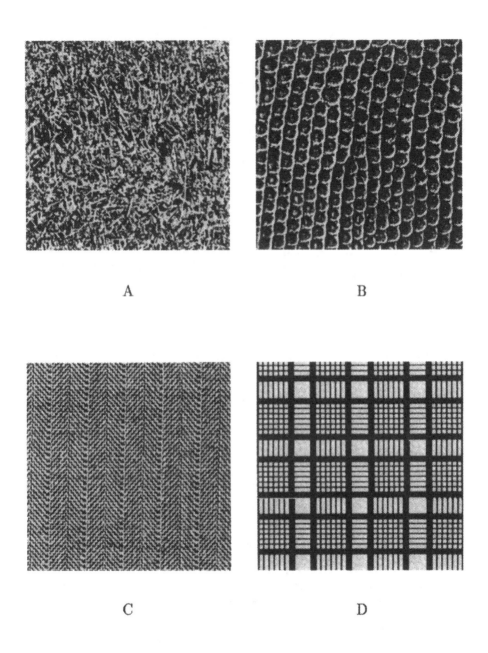

A

B

C

D

Figure 7.2: Four real textures: A, B, and C from Brodats (1966), and D from Hornung (1976).

Pattern A consists of many small texture elements evenly scattered in the image. It is analyzed statistically without regard to the texture elements. Pattern B consists of large texture elements scattered evenly in the image. It is analyzed structurally using the texture elements. Pattern C consists of many small texture elements which form local clusters in the image. The clusters are detected statistically by image segmentation without regard to the texture elements, and the pattern is analyzed structurally using the clusters. Pattern D consists of large texture elements which form local clusters in the image. The clusters are detected structurally by grouping, and the pattern is analyzed structurally using the clusters.

Figure 7.3 shows the flow of processing in the adaptive texture analysis system.

1) Statistical texture analysis: The system first computes the scalar texture properties of every input image by statistical texture analysis. They are used to roughly classify the image into the category which has the similar texture properties. If no competition occurs in image classification, the analysis may be stopped here. Otherwise, however, detailed structural texture analysis is necessary to classify the image into only one category. Thus, this step has a role of selecting as few candidate categories to be matched with the image in later structural texture analysis.

2) Intensity-based image segmentation: The system segments the image into regions of uniform intensity to define texture elements. If too many regions are extracted, however, the image is recognized to have *microtextures* in it like pattern A or C.

3) Texture-based image segmentation: If the system knows existence of microtextures in the image in step 2), it computes the scalar texture properties locally in the neighborhood at each point in the image by statistical texture analysis, and then segments the image into regions of the consistent texture properties.

If no more than one region is extracted, in other words, each point in the image has the consistent texture properties, then the system knows that the microtextures are scattered evenly in the image like pattern A. If there is no competition in image classification, the analysis may be stopped here. Otherwise, the uniform intensity regions extracted in step

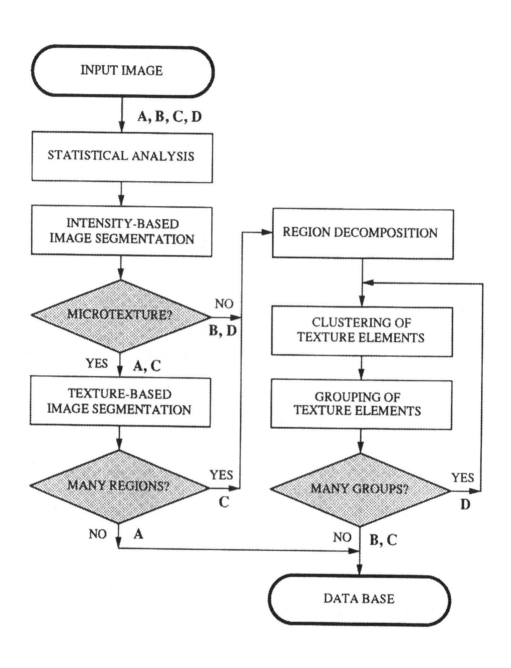

Figure 7.3: Adaptive texture analysis.

2) are used to define texture elements.

If many regions are extracted, on the other hand, the system knows that the microtextures clusters separately in the image like pattern B, and they are used to define texture elements.

4) Region decomposition: The system decomposes any region of complex shape, which has been extracted in step 2) or step 3), into subregions of simple shape. The resulting simple regions are defined as the atomic texture elements.

5) Clustering of texture elements: The system computes the properties of each texture element and the relative position vectors between texture elements, and then finds the sets of similar texture elements arranged by the same placement rule (defined by the consistent relative position vectors).

6) Grouping of texture elements: The system finds groups of texture elements in each set which are linked in the image space in a sense that there is a path between any pair of texture elements in a group by linking the neighbors of texture elements at the consistent relative positions.

If all the texture elements are linked into one group, the system knows that the texture elements are evenly scattered in the image like pattern B or C. If many groups of texture elements are detected, on the other hand, the system knows that texture elements form clusters in the image like pattern D. Each group is defined as a unit of analysis and it is used in the same way as an atomic texture element. Also, the units may be grouped into the larger units. Clustering and grouping is repeated until no larger unit is newly generated.

As a result, the system characterizes the given texture by the number of levels of groups, the number of groups in each level, the distributions of properties of texture elements in each group, and the placement rule of texture elements in each group and between groups.

7.2   Image Classification

There are two phases in image classification: the learning phase and the recognition phase. In the learning phase, the training images of which the categories are known are analyzed and the standard description of each category of texture is constructed which at least covers all the descriptions of the training images in the same category. In the recognition phase, every input image is analyzed to be classified into the proper category of which the standard description matches the best to the output description of the image. The process of each phase is not simple when analyzing textures structurally because the analysis is not uniform to every image. The problem to use structural analysis for image classification is that the groups of texture elements are not always detected in the same way even though the textures belong to the same category. For example, when the groups of texture elements are detected by recursively thresholding the histograms of the properties of texture elements, the same numbers of clusters are not always detected in the histograms.

7.2.1 Learning Phase

One way to solve the problem is to select the minmum number of clusters detected in the training images in the learning phase. Let the number of clusters of texture elements and the associated thresholds to sort the texture elements into the clusters be obtained in each histogram by analyzing the first training image. They are used as the standard numbers and the standard thresholds to analyze the following training image, respectively.

Consider a case that the number of clusters detected in a histogram is larger than the standard number analyzing another training image. The surplus thresholds which are farthest from the standard thresholds are discarded and the texture elements are sorted by using the remaining thresholds. And, the standard value of each corresponding threshold is refined to the average value. On the other hand, if fewer clusters are detected, the smaller number of clusters and the associated thresholds are used as the standard number and the standard thresholds to analyze the training images again from the first.

For example, Figure 7.4 shows the intensity histograms of texture elements of the first and the second image in the same category, respectively. The number of thresholds in the second histogram is larger by one than the number in the first histogram. In this case, the surplus threshold 21 in the second histogram which is farther from the threshold 32 in the first histogram is discarded, and the threshold 28 is used

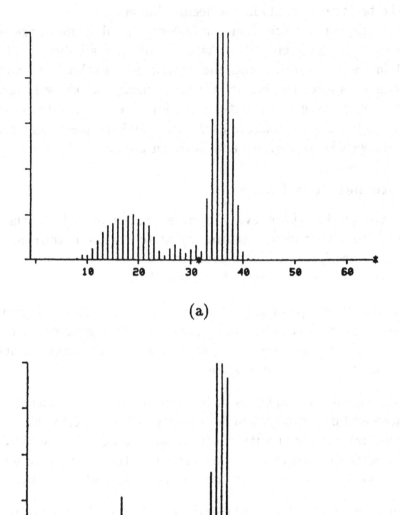

(a)

(b)

Figure 7.4: Determination of thresholds.

to sort the texture elements in the second image.

In this way, the texture elements in every training image are sorted into the same number clusters. As a result, the number of clusters detected in each histogram and the associated thresholds to sort the texture elements into the clusters are memorized with the standard description of each category of texture which includes the distributions of properties of texture elements in each group and the placement rules of texture elements in each group and between groups.

7.2.2 Recognition Phase

In the recognition phase, every input image is analyzed and classified into one of the categories of textures by matching the description with the standard decription of each category of texture. There are three kinds of items in the description to be matched.

1. Number of groups of texture elements: The number of groups of texture elements in the input image is first compared to that in each category of textures. Categories with different number of groups of texture elements are discarded.

2. Distributions of properties of texture elements: The distributions (mean and deviation) of each property of texture elements in each group are compared with those of each category. The similarity between two distributions is evaluated by the common area of the two distributions assuming that they are normal distributions.

3. Placement rules of texture elements: The distributions of relative positions between texture elements in each group and between groups are compared with those of each category in the same way as the distributions of properties of texture elements.

The input image is classified into the category with which the sum of similarity is the largest. If the best similarity is too low, however, the decision is put off. The low similarity may be caused by wrong thresholds to sort the texture elements into clusters in the input image. In such a case, the analysis is retried for each category by matching the thresholds in the input image to the standard thresholds in the

Table 7.1: Classification accuracy on test samples

ID. NO.	D3	D9	D15	D20	D28	D32	D33	D34	D49	D67	D68	D84	D93	D103	D109	D111	CLASSIFICATION ACCURACY (%)
D 3	12																100
D 9		12															100
D 15			12														100
D 20				12													100
D 28					12												100
D 32						7	4									1	58
D 33							11									1	92
D 34							1	11									92
D 49									12								100
D 67										10			2				83
D 68											12						100
D 84												12					100
D 93													12				100
D103														11	1		92
D109														1	11		92
D111			1												3	8	67

category. If the number of clusters detected in a histogram of the input image is smaller than the standard number, the texture is not classified into the category. On the other hand, if the number is larger, texture elements are sorted without using surplus thresholds and the revised description are computed. Lastly, the input image is classified again into the category with the largest similarity.

7.2.3 Experimental Results

For our experiments we selected 16 textures with comparatively small texture elements from the Brodat's photographic album (1966). Their identifying numbers are D3, D9, D15, D20, D28, D32, D33, D34, D49, D67, D68, D84, D93, D103, D109, and D111. Each image is digitized into 256×256 pixels and quantized into 64 intensity levels. We divided the image into 4×4 subimages with 64×64 pixels and used 4 diagonal ones for training and the remaining 12 ones for test.

We achieved a classification accuracy of 100 % on the training samples and the results on the test samples are shown in Table 7.1. Since most errors are due to the small size of the images, the performance would be greatly improved if we use larger images. In the case of D111, for example, the grains of texture elements vary greatly in different samples. D32 and D33 are both textures of cork but slightly differ in scale. We can compare the scale of image by the size of texture elements, but there was no significant differences because of the resolution of the images.

7.3   Analysis by Synthesis

Texture images are generated from the descriptions of the original texture images given by structural analysis. The purpose of the texture generation is to evaluate the description. In the description of a texture are the number of groups of texture elements, the detailed shape of typical texture element in each group, the distributions of intensity and shape properties of texture elements in each group, and the placement rules of texture elements (the distributions of relative positions between texture elements) in each group and between groups. The reconstructed textures will have at least the same description as those of the original texture.

7.3.1   Generation of Texture Elements

Texture elements are generated by modifying the typical texture element in each group according to the distributions of properties of intensity and shape of texture elements in the group. Let (μ_{ij}, σ_{ij}) be the mean and the deviation of the jth property of texture elements in the ith group. Texture elements in the ith group are generated by modifying the typical texture element so that the distribution of the jth property of generated texture elements becomes the normal distribution $N(\mu_{ij}, \sigma_{ij})$, using a normal random number generator. The shape properties are used to modify the shape of typical texture elements. For example, size is used for scaling, magnitude of elongation for compression, and direction of elongation for rotation. The intensity of each texture element is given according to the distribution of intensity.

7.3.2 Placement of Texture Elements

Texture elements generated are arranged in a white image according to the placement rules. Let Φ_i be the placement rule of the ith group of texture elements which are the distributions of the first nearest and the second nearest relative position vectors between texture elements in the group, and Φ_{ik} $(i \neq k)$ be the placement rule between the ith group and the kth group of texture elements which are the distribution of the nearest relative position vectors between pairs of texture elements, one in the ith group and the other in the kth group.

First, if there is more than one group of texture elements, the order to place texture elements is determined according to the regularity of each group of texture elements. The regularity is evaluated by the deviation of the first vector of the placement rule, that is, the smaller is the more regular. The most regular group is called the *dominant group* in which the texture elements are first placed in the image. The texture elements in other groups are then placed relatively to the texture elements in the dominant group. The most irregular group of texture elements forms the background of the image.

Let the ith group be the dominant group now. First, the periodic positions are computed using the means of the first and second vectors in the placement rule Φ_i, as represented by the white points in Figure 7.5. Next, these preliminary positions are modified using the deviations of the first and second vectors, as represented by the black points in Figure 7.5. These black points are the final positions at which generated texture elements are placed.

The positions of texture elements in the kth group are determined according to the placement rules Φ_k and Φ_{ik}. First, a base position is fixed relatively to the position of one texture element in the dominant group using the mean of the vector in the placement rule Φ_{ik}. Other positions are determined from the base position in the same way as those of the dominant group of texture elements using the placement rule Φ_k.

After all the texture elements except for the last group of texture elements have been placed, blank spaces remain in the image. They correspond to the ground of the image. They are filled with intensity values according to the intensity distribution of the last group of texture elements.

111

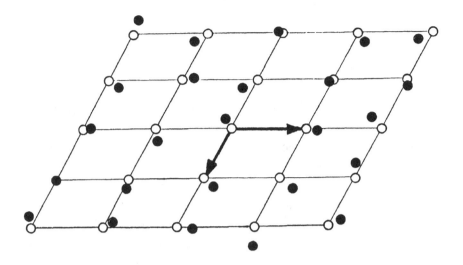

Figure 7.5: Placement of texture elements.

For example, Figure 7.6 shows the original images and the reconstructed images.

7.3.3 Generative Models of Textures

The study of modeling textures is important as the base for both texture synthesis and texture analysis. Since it is possible to control the characteristics of the texture generated by the model, it is useful in the theoretical analysis of textures or in the psychological experiments. The proposed structural analysis is based on *structural model* of textures. There are some other generative models of textures as follows

When raster scanning the image, the sequence of intensities can be regarded as *Marcov chains*. The transition probabilities of the mth-order Marcov chain specifies the $(m+1)$th-order statistics of textures. Julesz (1975) and Pratt et al. (1978) have generated textures with different second-order statistics by first-order Marcov chains, and Gagalowicz (1979) has generated textures with different higher-order statistics by higher-order Marcov chains. They are used for psychological experiments.

112

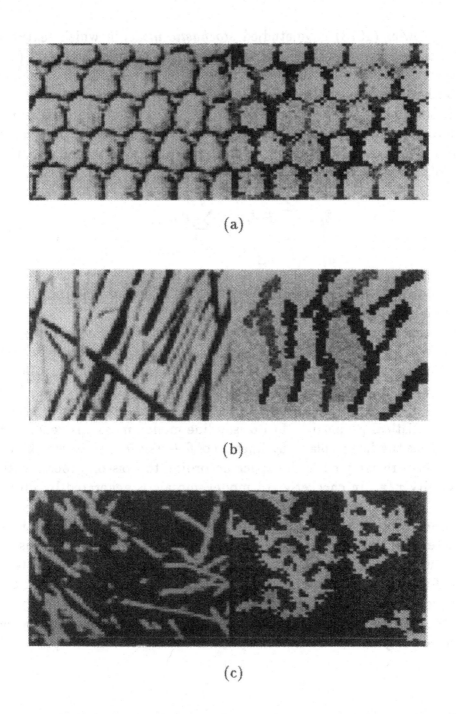

(a)

(b)

(c)

Figure 7.6: Original images (left) and reconstructed images (right)

Gagalowicz (1978) have studied *stochastic model* in which textures are generated from a white noise $B(k)$ through transfer function $H(z)$. Pratt et al. (1978) and Hassher et al. (1978) have taken the similar approaches to synthesize textures.

McCormick et al. (1974) and Tou et al. (1976) have used *time series model* to reconstruct textures. Let $\{Z_t\}$ denote a time series. The value of Z_t is generated (estimated) by the terms of autoregression and the terms of moving average:

$$\hat{Z}_t = \sum_{i=1}^{p} \phi_i Z_{t-i} + \sum_{i=1}^{q} \theta_i a_{t-i}$$

$$a_t = \hat{Z}_t - Z_t$$

(7.1)

where ϕ_i and θ_i are coefficients of which the values are determined by least square method.

Schachter et al. (1978) has proposed *random mosaic model* to generate and analyze cellular textures. In the processes to generate mosaic are occupancy model, Johnson-Mehl model, Poisson line model, rotated checkerboard model, bobming model, etc. Among them, Poisson line model and checkerboard model can be used to analyze textures based on the statistical properties. In Poisson line model, mosaic is generated by dividing the image plane by lines $x \cos \theta + y \sin \theta = \rho$, of which the parameters are sampled in θ-ρ space according to Poisson process with the density τ/π. In checkerboard model, mosaic is generated by filling the image plane randomly with square cells of side length b.

Mandelbrot (1982) has proposed *fractal* which is a set of functions characterized by non-integer fractal dimension to describe patterns in the natural world like terrains, clouds, trees, etc. The fractal pattern is a self-similar pattern even in a statistical sense which has a infinitely recursive structure. Pentland (1984) has applied the fractal theory to analyze textures and demonstrated that there is a correlation between perceived roughness and fractal dimension of textures.

Chapter 8

Object Recognition

The ultimate object of image analysis by computer, which includes texture analysis, image segmentation, shape analysis, and grouping so far discussed, is to automatically recognize objects in the images. The difficulty of recognition is in variety of analysis; different programs are needed to analyze different kinds of images. Generally speaking, the computer must learn about objects in images before recognition. Learning is currently the main issue in artificial intelligence. The learning strategies are ordered from the easiest one to the most difficult one according to the amount of inference involved: 1) learning by being programmed, 2) learning by being told, 3) learning by seeing samples, and 4) learning by discovery (Winston, 1977).

The first strategy has been the conventional way in which image processing experts have coded a special program whenever a new kind of image is analyzed, developing new algorithms. Now that there are many useful algorithms and even commercialized subroutine libraries are available (Tamura et al., 1983), it is necessary to develop a system with higher learning strategies in which even a user who is not an image processing expert, for example, a physician, can easily create a program which analyzes a specific kind of image, for example, medical images.

The current image processors work by commands, the second strategy. However, it is still difficult to use them if the user doesn't know the specifications of programs. The expert systems for image processing teach the user how to select proper programs by production rules

(Matsuyama, 1986). However, they still belong to the second strategy.

This chapter presents an object recognition system IMARS (Interactive Modeling and Automatic Recognition System) which has been developed to support the user, who has little knowledge of any image processing algorithms, to create a program to recognize specified objects in a specified kind of image, by the third strategy, mainly depending on visual information (Tomita, 1983; 1988).

8.1   Overview of System

There are at least two steps in recognizing objects in a given image, as shown in Figure 8.1:

1) extraction of units of objects to be recognized in the image, and then

2) identification of the units with known objects based on the properties of the units.

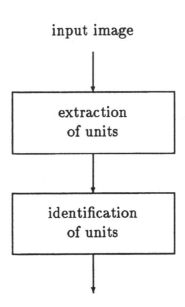

Figure 8.1: Two steps in object recognition.

It is assumed that any object in any image is represented by one of the units: image itself (in this case only, the first step is not needed), region, medial-axis, line, vertex, or set (group) of units. For example, when diagnosing a stomach X-ray image in Figure 8.2(a), a region is recognized to be *ventriculus*, a line to be *curvatura ventriculi major*, and a vertex to be *angulus ventriculi*, as shown in Figure 8.2(b).

Each step in the "extraction–identification" cycle is highly dependent on both objects and images. It is different from object to object and from image to image. It might be necessary to repeat these cycles to recognize complex objects. In principle, we must write a new special program whenever we analyze a new kind of image. It is possible, however, to avoid such programming waste by separating image processing algorithms from image dependent information.

In IMARS, image processing algorithms are classified based on the functions and the corresponding programs are moduled in the library. Each program has some parameters and the values of these parameters depend on the objects and the images. Then, the problem is how to select a set of useful programs and how to determine the values of parameters.

There are two modes in implementing IMARS: interactive mode and automatic mode. First, in the interactive mode, the user tests effectiveness of given programs interactively by trial-and-error using training images. If he succeeds to recognize objects in the image, then he can store the program numbers, the associated parameters, and the properties of objects in the data-base called *model*. The model is refined in a bottom-up way by analyzing as many training images in the same category. Once the model is completed, the system can automatically analyze images in the same category and recognize the expected objects in a top-down way driven by the model in the automatic mode.

Each category of images will have its own model. Thus, the system can recognize many kinds of images simply by retrieving the corresponding models.

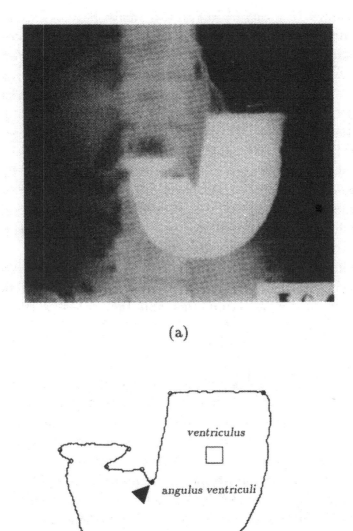

(a)

ventriculus

angulus ventriculi

curvatura ventriculi major

(b)

Figure 8.2: (a) A stomach X-ray image; (b) Recognition of the portions of the stomach.

8.2   System Configuration

The system is composed of program library, image processor, image descriptor, model, and interface, as shown in Figure 8.3. The role of each block is as follows:

- Program Library: Image processing programs are stored. Programs are called with the specified values of the parameters by the user or the model.

- Image Processor: It consists of image memories. Each program uses them for input images, label images, work images, and output images.

- Image Descriptor: It is a short-term memory which temporarily records the program numbers and the associated parameters used to extract units in the current image, the properties of these units, and the result of identification of these units. The records are used to create and refine the model, and to identify units with the objects in the model.

- Model: It is the data-base which stores names of objects to be recognized, the program numbers and the associated parameters used to extract the objects in the image, and the properties of the objects.

- Interface: It exchanges information between user, program library, image descriptor, and model. It displays the input image, the units extracted, the result of identification of these units, and other useful information such as contents in the model. The user can communicate with the system by means of the keyboard and the graphic display with cursors.

8.3   Extraction of Units

In the types of units to represent objects in the image are image, region, boundary, medial axis, vertex, and group of units. First, the user must determine the type of the unit of each object which he wants

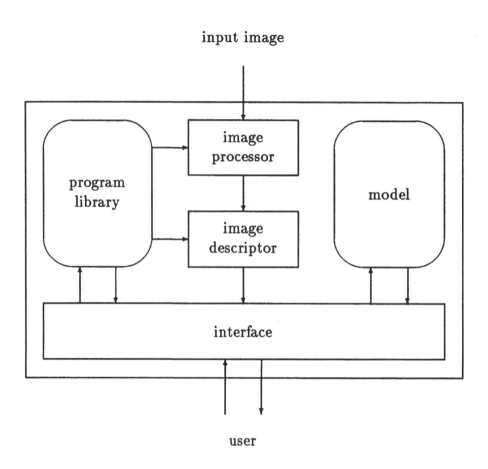

Figure 8.3: System configuration.

to recognize, and then select the program to extract them. Units are extracted by either segmenting, merging, or grouping units. For example, segmenting a region produces more than one smaller region, merging adjacent regions produces one larger region, and grouping regions produces one cluster (group) as a set of petals forms a flower, as shown in Figure 8.4.

The corresponding program is composed of a fixed sequence of program modules. For example, the edge detection program, which segments the image (or region) into regions (or subregions), consists of

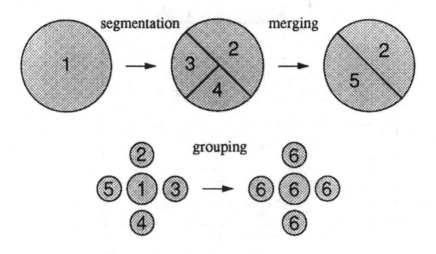

Figure 8.4: Three types of region extraction.

program modules for computing local textures, computing texture gradients, and thinning–thresholding–extending edges. The user need not select each module in the program but he must set the values of the parameters associated with each module, for example, the regions processed, the texture property used, the size of the gradient operator, and the threshold to suppress weak edges, as shown in Figure 8.5. The system asks the user to set the values of the parameters by showing the menus with default values. The user can answer to it by means of the keyboard or the cursors. When he set a threshold to suppress weak edges, for example, he may give a numeric value directly or indicate some points in the image by cursurs so that the threshold is set lower than the edge strengths there.

IMARS offers the latter type of procedure to set parameter values for as many programs as possible since it is a difficult problem even for image processing experts to detemine continuous values of parameters. Since the result of processing is displayed at each stage, he can easily evaluate it. If the result is unsatisfactory for him, he can backtrack to the previous stage and retry the program by changing the values of

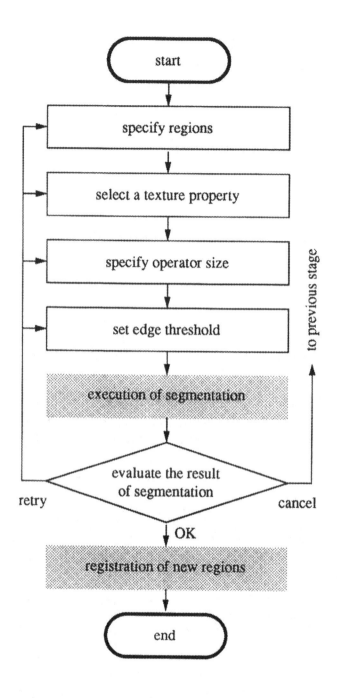

Figure 8.5: Setting parameters in edge detection program.

parameters until he gains a satisfactory result. He may select another program, for example, region thresholding instead of edge detection to extract regions, if the current program does not work well.

8.4   Properties of Units

There are two types of properties the system must compute for units of objects in the image: the properties of unit itself and the relations between units. In the properties of region itself, for example, are intensity (or texture), shape, and position. In the relations between regions are difference of the properties of region itself (relative intensity, relative position, relative orientation, etc.), adjacency, and symmetry. The system computes all the properties of unit itself for every unit extracted in the image. As for relations, on the other hand, the system computes only a specified relation between a specified pair of units.

If the number of objects is a few and the number of relations is also a few as in Winston's block world (Winston, 1970), it may be possible to compute all relations between all pairs of objects. In the real world, however, there are many objects including objects which are not necessary to be recognized and also many kinds of relations can be defined. It is wasteful to compute all relations between all pairs of objects. The user should define a specified relation between a specified pair of objects only when the system cannot recognize the objects based only on the properties of each object itself. Then, the system will first find candidate units of objects based only on the properties of objects themselves, and then test the relation only for pairs of the candidate units.

8.5   Structure of Model

The model is represented by a graph. For example, the model which analyzes cranial CT (Computer Tomography) images is shown in Figure 8.6.

- A node represents a kind of object to be recognized in the image. Particularly, a root-node (HEAD) represents a category of image (cranial CT image). Each node has value slots for its name, the

123

type of the unit in the image, and the properties of the unit. The properties have a range of values. The properties useful for identifying the object will have a narrow range of values. Whereas, less important properties will have a wider range of values.

- A solid link represents the *parent-and-child* relationship. Child objects are extracted by segmenting, merging, or grouping parent objects. The programs used and the associated parameters are stored at nodes of parent objects. The parameters also have a range of values. Each parameter is adjusted within the range to extract child objects.

- A dotted link represents *OR* relationship. One of the linked objects will be recognized. For example, $B(1), \cdots, B(6)$ correspond to brains of different cross sections of a head. The OR relationship may be used to represent an object which cannot be defined by a single node, that is, in a unique way because of its variety. When more than one kind of image are analyzed by one model (for image classification), multiple root nodes are defined in the model. They have OR relationship from the beginning.

- A white node signifies that the object is always recognized, whereas a black node signifies a *free* object which is not always recognized. For example, abnormalities (L and H) in cranial CT images are free objects which do not always exist.

8.6 Building Model

In the interactive mode, the user teaches the system what objects he wants to recognize, how they are extracted in the image, what properties should be measured, and what property values they have, analyzing training images in a bottom-up way. This procedure is called *modeling*. The system stores what it learns in the model.

8.6.1 Initial Model

First of all, the user gives the name of the category of images to be analyzed. Then, the system creates the root node with its name in

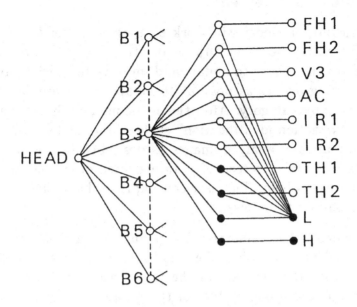

Figure 8.6: Model to recognize cranial CT images.

the model and computes the texture properties for rough classification before detailed analysis. Next, the user tries to extract units of objects in the image by selecting one of the given programs. He will adjust the values of the associated parameters several times by trial-and-error until the desired units are extracted properly. When he succeeds in extracting the units of objects, he indicates them by a cursor and gives their names. Then, the system creates the new nodes linked to each parent node in the model. At the parent node, the program number used and the values of the associated parameters are stored. At the new node, the object name and the properties of the corresponding unit are stored.

When it is necessary to further extract child objects from the detected objects, the user indicates the units to the system by pointing to them or using their names. Then, the system masks the specified units for the user to be able to analyze only the masked units. Resulting child (descendant) units are registered in the same way as before. This procedure is repeated until all the descendant objects are defined in the model.

125

8.6.2 Model Refinement

Since the initial model will work properly only for the first image and also there may be other objects in other images which should be recognized, it is necessary to update and improve the initial model using as many training images as possible to be able to be applied to all the images in the same category. After the initial model is built, the user can test every "extraction-identification" cycle of the model by making the system automatically analyze images as described in the next section. When the system fails to recognize an expected object at any stage, it demands the user's help. The user can improve the model according to the cause of failure as follows.

- Refinement of Parameters: If no unit of an expected object is extracted properly, then the value ranges of the parameters associated with the program in the model are insufficient. The user must find the proper values of the parameters by trial and error until the unit is extracted in the same way as when making the initial model. And, indicating the resulting unit by a cursor and giving the object name revise the maximum or minimum values of the parameters in the model. After this the parameter values will be adjusted within the new range.

- Refinement of Properties: If a unit of an expected object is extracted but is not identified, then at least one of the value ranges of properties of the object in the model are insufficient. The user must indicate the unit by a cursor and give the object name. Then the maximum or minimum values of the properties of the object in the model are revised and the unit will be identified properly.

- Refinement of Relations: If an expected relation is not found for a pair of objects, the value range of the relational property in the model is insufficient. The user must indicate the pair of units by a cursor and give the index of the relationship. Then the maximum or minimum values of the relational property in the model is revised and the missing relation comes into existence.

- Declaration of OR Relationship: If there is not a registered object (let it be A) but another object (let it be B) instead. The user must register B in the model and inform the system that A and B have OR relationship.

- Declaration of Free Object: If there is an object which does not always exist, the user must inform the system that it is a *free* object.

If there is an identification error, for example, object A is mistook for object B, then the two objects have a common area in the property space. The user must take one of the following actions to remove the common area.

- If there is an relational property between one object (let it be A) and another object C but between B and C, the user must indicate the pair of corresponding units by a cursor and give the index of the relationship. Then the two confusing objects will be distiguished based on the relation.

- If one object is preferential, then the user can define OR relationship between the two confusing objects and determine the order of priority.

- Otherwise, it is impossible to distinguish the two confusing objects since the definition of at least one object is wrong. it is necessary to remove information related to the wrong object in the model and redefine it in another way.

Modeling is repeated until all the training images are analyzed properly by the model. Generally, there may be a large burden on the user first but it decreases as the system grows since the amount of refinement decreases. A flexible model will be built by making both the program parameters and the object properties have a range of values, and by making the objects have OR relationships.

8.7 Automatic Recognition

Once the model is completed, the system can automatically analyze images in the same category in a top-down way driven by the model.

127

However, there is a problem that parameters in the program to extract objects have a set range. The system must select the best value of each parameter within the range so that it could extract the expected objects properly. It is impossible to examine all the values because of its continuity. Let a parameter have the maximum value p_{max} and the minimum value p_{min}. IMARS tries p_{max}, $(p_{max} + p_{min})/2$, or p_{min} for the parameter. If there are n parameters in a program, the system tries a maximum of 3^n cases until it finds all expected objects by this process.

Parent objects will be identified first. However, it is a temporary decision until all their descendant objects are identified properly. Descendant objects are usually detailed descriptive components of the parent objects. When the system fails to find any expected object at one "extraction–identification" cycle even by adjusting the values of the parameters, it backtracks to the previous level of identification of the parent unit. It tries another interpretation. Backtracking is repeated until the most descendant objects are identified properly. However, if the system fails to find any expected object even by backtracking to the root node, it demands the user's help. The cause of the failure is insufficiency of the model or the input image in a different category. In the former case, the user must teach the system how to analyze the image in the interactive mode and improve the model.

8.8   Experimental Results

We first used the system to recognize cerebrospinal fluid regions and abnormalities in cranial CT images (Ramsey, 1981). Cranial CT images are input from the MTs of a GE CT/T Scanner System. The image size is 320×320 points and the intensity level is 11 bits. Figure 8.7(a) shows an example of the cranial CT images. Figure 8.7(b) shows the intensities of the image along a horizontal line. We can see that there is a high contrast between the skull and the brain, and that there are textures in the brain. Since the shape and the position of the classified portions differ depending on individuals and ages, a flexible model is necessary to recognize them. The model to analyze the cranial CT images has been shown in Figure 8.5. Each object in cranial CT images is represented by a region.

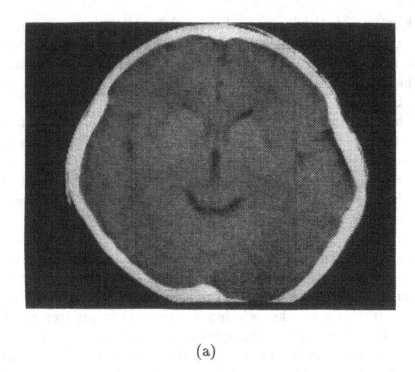

(a)

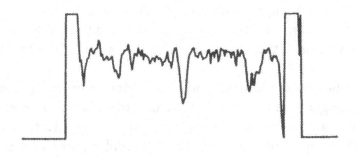

(b)

Figure 8.7: (a) Cranial CT image; (b) Intensities of the image along a horizontal line.

1) First, given the image in Figure 8.7, the system computes the statistical texture properties of the image and then identify it with (a candidate for) HEAD, the cranial CT image, based on the values.

If the system fails to recognize HEAD during the test of the model, it gives the message "NO MODEL MATCHES." In this case, the user inputs the code to register an object in the model. As the system demands the name of the input image, the user types "HEAD." Then, the system refines the ranges of properties of HEAD in the model.

2) Next, the system segment the image into regions of uniform intensity by the edge detection method, as shown in Figure 8.8(a), computes the properties of each region, and then identifies the region in Figure 8.8(b) with brain B(3).

If the system fails to recognize B(3) during the test of the model, it gives the message "B(3) IS NOT FOUND." If the corresponding region is extracted properly, the ranges of some properties in the model are not sufficient. In this case, the user inputs the code to register an object in the model. When the cursur appears in the display, the user moves the cursor in the corresponding region and clicks there. As the system demands the name of the region, the user types "B(3)."

If the corresponding region is not extracted properly, on the other hand, the ranges of some parameters in the model are not sufficient. In this case, the user inputs the code to backtrack to the stage before segmentation. Then, the user segments the image interactively until the proper region is extracted, and makes the system identify the new regions.

3) Next, the system segments only the region of brain B(3) into sub-regions of homogeneous texture by the same edge detection method but with different parameters, as shown in Figure 8.9(a). And, the system recognizes the candidates for cerebrospinal fluid regions and abnormal low density regions based on the properties of each region, as shown in Figure 8.9(b).

4) Finally, if there are adjacent regions which are identified with the same object, the system merge them into one region. Then, the system recognizes the normal cerebrospinal fluid regions (FH(1), IR(1), etc.) and one abnormal low density region (L), as shown in Figure 8.10.

Here, the system uses the symmetry relation to distinguish between

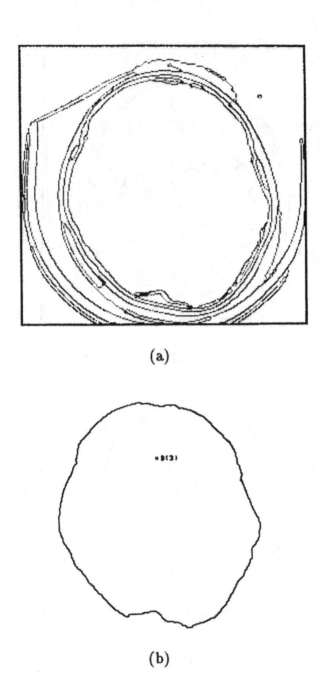

(a)

(b)

Figure 8.8: First cycle of recognition: (a) Regions extracted by edge detection method; (b) Identification of brain B(3).

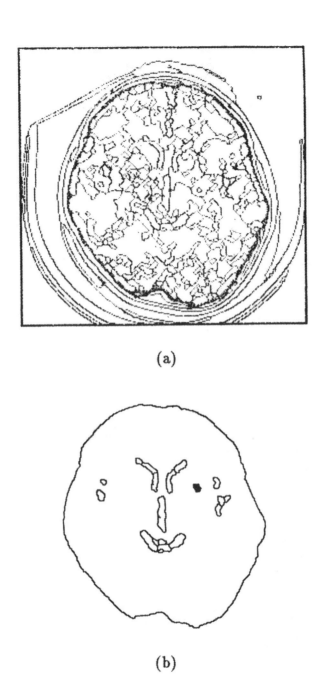

(a)

(b)

Figure 8.9: Second cycle of recognition: (a) Regions extracted by edge detection method; (b) Identification of candidates for brain components.

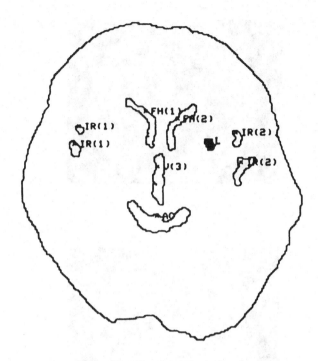

Figure 8.10: Final result of recognition.

normal cerebrospinal fluid regions and abnormal low density regions because they could have the same region properties. Pairs of normal cerebrospinal fluid regions (for example, FH(1) and FH(2)) have a symmetrical axis about the vertical center line of the brain, whereas abnormal regions usually do not have it.

If the relation is not satisfactory during the test of the model, however, the system gives the message "FH(1) IS MISSING FH(2) WITH SYMMETRY." In this case, the user inputs the code to register a relation in the model. As the system gives the menu of relations, the user selects SYMMETRY among them and points to the regions corresponding to FH(1) and FH(2) by the cursur, respectively.

And, when all the objects in the model are recognized properly, the system gives the message "SUCCEEDED."

Other recognition results are shown in Figure 8.11. The low density abnormalities (L) are detected in Figure 8.11(a) and 11(b), and the high density abnormality (H) is detected in Figure 8.11(c).

133

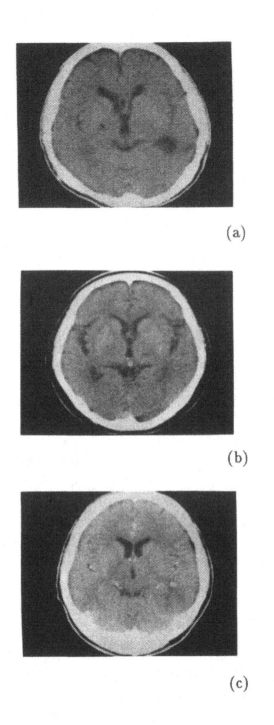

(a)

(b)

(c)

Figure 8.11: Results of analysis of cranial CT images.

8.9 Summary

IMARS can be used even when the user does not know exactly what model can analyze his data since a model is built in a bottom-up way. The user himself learns about the objects with the system while he interactively constructs the model. The necessary condition is that necessary programs are provided in the library. In IMARS, programs are classified based on functions and one program is prepared for one function. The most general algorithm is used among algorithms which work almost in the same way. For example, though many kinds of edge detectors have been proposed, only one normal gradient operator whose size is variable is being used. As a result, it is necessary for the user to think only functions, and it is rare that the user get puzzled which program to use. If a necessary function is lacking, it can be easily discovered while using the system. It is necessary to test the system using as many kinds of images to find lacking functions. It is also necessary to decrease burdens on the user more by automating even the current options. Such a system must have the last learnig strategy, learning by discovery.

This kind of pattern recognition system will be the bridge between the real world and the current expert systems or symbolic reasonong systems. The current reasoning systems are processing "words" which are defined by a human. Those systems don't know real objects which correspond to the words. On the other hand, the pattern recognition system produces words, that is, symbolic representations of objects in the real world. Combining the pattern recognition system and the symbolic reasoning system will produce more powerful expert systems which will decrease human's burdens.

Chapter 9

Shape from Texture

When an image is a projection of a three-dimensional (3D) scene, some 3D forms in the scene can be estimated from 2D features extracted in the image. Horn (1977) first proposed a method to estimate the surface orientation from the shading in the image. It is called *shape-from-shading*. When we observe the surface covered with texture even by a single eye, we perceive the inclination of the surface because the texture on the surface is apparently distorted according to the inclination. The role of texture as a basis for the recovery of surface orientation was first investigated by Gibson (1950), and the method to estimate the surface orientation from the apparent texture distortion is called *shape-from-texture.*

Texture distortions under perspective projection are caused by the gradient of the surface, the angle between the line of sight and the image plane, and the distance from the view point to the surface. Then, the problem is how to know these factors from the image. Since the problem is under constraints, some additional assumtions are necessary. Accordingly, various algorithms have been proposed to solve this problem. Table 9.1 classifies shape-from-texture algorithms based on the surface cue—texture gradient, converging lines, normalized texture property map, or shape distortion, the surface type—planar or curved, a priori knowledge about the original texture, the projection type—orthographic, perspective, or spherical, the level of texture analysis—statistical or structural, the unit of texture analysis, and the property

Table 9.1: Taxonomy of shape-from-texture.

section	author (year)	surface cue	surface type	original texture
9.1.1	Bajcsy (1976)	texture-gradient	planar	unknown
9.1.2	Ohta (1981)	texture-gradient	planar	unknown
9.1.3	Aloimonos (1985)	texture-gradient	planar	uniform-density
9.2.1	Nakatani (1980)	converging-lines	planar	parallel-lines
9.2.2	Kender (1979a)	converging-lines	planar	parallel-lines
9.3.1	Kender (1979b)	NTPM*	planar	known
9.3.2	Ikeuchi (1981)	NTPM*	curved	known
9.4.1	Witkin (1981)	shape-distortion	planar	isotropy
9.4.2	Walker (1984)	shape-distortion	curved	unknown

*normalized texture property map

Table 9.1: (continue)

projection type	level of analysis	unit of analysis	property of unit(s)
perspective	statistical	wave	length
perspective	structural	region	area
perspective	statistical structural	edge elem. region	density
perspective	statistical	edge elem.	direction
perspective	statistical	edge elem.	direction
orthographic	structural	line etc.	length etc.
spherical	structural	region	axes
orthographic	statistical	edge elem.	direction
orthographic	structural	region	shape

of the unit. The assumption common to all the algorithms is that the surface is smooth and is covered with the homogeneous texture. This chapter overviews these techniques for shape-from-texture.

9.1 Texture Gradient

Texture gradient, by which Rosenfeld (1975) means the rate and direction of maximum change of texture coarseness across the surface, is the most primary cue in determining the orientation of the surface relative to the observer's line of sight.

9.1.1 Bajcsy and Lieberman (1976)

They used a gradient in the size of texture elements as a depth cue for the longitudinal ground surface. The geometric model is shown in Figure 9.1. Let l_i be the observed texture element size at Y_i in the image. The texture elements are all assumed to be of the same size t. Y_A, Y_B, and Y_C are the y values of the projection of points A, B, and C in the image. A projection function $P(y)$, which indicates how distance s on the ground changes with distance y in the image, is represented by

$$P(y) = \frac{1}{k_1}\frac{ds}{dy} \qquad (9.1)$$

where k_1 is a constant dependent on the geometric parameters of the camera model (focal length, height above the ground, field of view angle). Then the distance in the scene is obtained by

$$s = k_1 \int P(y)dy \qquad (9.2)$$

A form of $P(y)$ is approximated by fitting a functional curve to the observed values:

$$P(Y_i) = k_2\frac{t}{l_i} \qquad (9.3)$$

where k_2 is another constant. Then the relative distance is compared by

$$\frac{\overline{AC}}{\overline{AB}} = \frac{k_1 k_2 t \int_{Y_A}^{Y_C} P(y)dy}{k_1 k_2 t \int_{Y_A}^{Y_B} P(y)dy} = \frac{\int_{Y_A}^{Y_C} P(y)dy}{\int_{Y_A}^{Y_B} P(y)dy} \qquad (9.4)$$

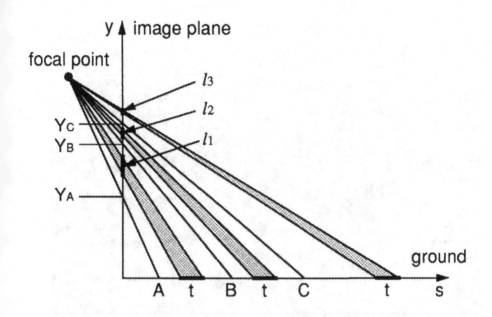

Figure 9.1: A geometric model for interpreting the texture gradient (from Bajcsy and Lieberman, 1976).

This means that the depth information is acquired through the gradient function $P(y)$ without knowing the constants of the camera model and the original texture element size t. The observed texture element size l_i is measured using the Fourier power spectrum in local image windows (see Section 2.2.3), as shown in Figure 9.2.

9.1.2 Ohta, Maenobu, and Sakai (1981)

They estimated the orientation of a planar surface from the rate of change of area of texture elements in the image. The geometric model is shown in Figure 9.3. Texture elements are on a planar surface $z = px + qy + c$. The center of gravity G of a texture element is projected at (X, Y) in the image plane $z = -1$. The z coordinate of the center of gravity is given by

$$d = \frac{c}{1 - pX - qY} \tag{9.5}$$

141

(a)

							Top of image (background)										Mean wavelength
12	21	21	18	18	21	18	21	21	21	21	21	21	21	21	21	21	20
26	26	26	26	26	26	26	26	21	26	26	26	26	26	26	26	26	26
32	32	32	32	32	32	32	32	32	26	32	32	32	32	32	32	32	32
32	32	32	32	32	32	32	32	32	32	32	32	32	32	32	32	32	32
43	43	43	43	43	32	43	43	43	43	43	43	43	43	43	43	64	44

Bottom of image (foreground)

(b)

Figure 9.2: (a) An ocean view; (b) Texture wavelengths in horizontal direction of ocean waves texture (wavelengths in the chart correspond to prefered frequencies in 128^2 windows) (from Bajcsy and Lieberman, 1976).

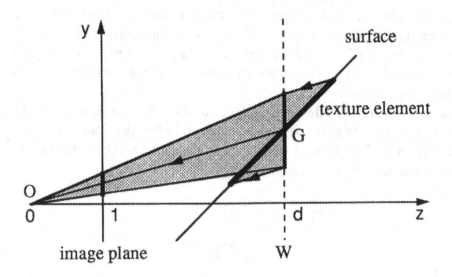

Figure 9.3: Approximation of perspective projection (from Ohta, Maenobu, Sakai, 1981).

If the size of each texture element on the surface is small, the perspective projection can be approximated by the following two steps of transformations.

1) The texture element is projected onto a plane W which is parallel to the image plane and includes the center of gravity G. The projecting rays are parallel to the central projecting ray OG. This transformation is given by an affine transformation:

$$T_1 = \frac{1}{\sqrt{p^2+1}} \begin{pmatrix} 1-pX & \dfrac{-q(p+X)}{\sqrt{p^2+q^2+1}} \\[2mm] -pY & \dfrac{p^2-qY+1}{\sqrt{p^2+q^2+1}} \end{pmatrix} \tag{9.6}$$

2) The image on the plane W is perspectively projected onto the image plane. This transformation is just a reduction of scale:

$$T_2 = \frac{1}{d} \tag{9.7}$$

143

The following operations are for obtaining the vanishing line in the image, the intersection of the image plane with the planar surface. The equation of the vanishing line is given by $px+qy+1$. When the vanishing line is obtained in the image, we can know the gradient (p,q) of the corresponding surface.

It is known that the determinant of an affine matrix is equal to the ratio of the areas of the two patterns before and after the transformation. Specifically, if S^* be the area of a texture element on the surface and S the projected area in the image, then we have:

$$\frac{S}{S^*} = \det T_2 T_1$$
$$= \frac{1}{d^3} \frac{c}{\sqrt{p^2 + q^2 + 1}} \qquad (9.8)$$

This means that the area of a texture element in the image is inversely proportional to the cubic of the distance of its center of gravity.

The distance from the center of gravity of the texture element to the vanishing line in the image is given by

$$h = \frac{1 - pX - qY}{\sqrt{p^2 + q^2}}$$
$$= \frac{1}{d} \frac{c}{\sqrt{p^2 + q^2}} \qquad (9.9)$$

This means that h is inversely proportional to d.

Let S_1 and S_2 be the areas of a pair of texture elements, h_1 and h_2 be the distances from each center of gravity to the vanishing line, and f_1 and f_2 be the distances from each center of gravity to the vanishing line along the line through each center of gravity, respectively, as shown Figure 9.4. From equations (9.8) and (9.9), we have the following equation.

$$\frac{f_1}{f_2} = \frac{h_2}{h_1} = \frac{d_1}{d_2} = \frac{\sqrt[3]{S_1}}{\sqrt[3]{S_2}} \qquad (9.10)$$

From this equation, one vanishing point (a point on the vanishing line) can be obtained since S_1, S_2, and $f_1 - f_2$ are known in the image. When there are n texture elements in the image, $_nC_2$ vanishing points can be obtained in the same way from every pair of texture elements,

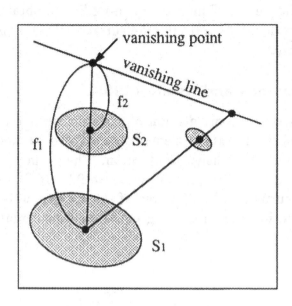

Figure 9.4: Vanishing point from pairs of texture elements (from Ohta, Maenobu, Sakai, 1981).

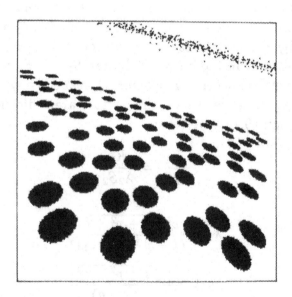

Figure 9.5: An artificially generated texture (determined vanishing points are dotted) (from Ohta, Maenobu, Sakai, 1981).

as shown in Figure 9.5. Then, a vanishing line is obtained by fitting a straight line to them. As a result, the surface gradient (p, q) is obtained from the line equation.

9.1.3 Aloimonos and Swain (1985)

They estimated the orientation of a planar surface from the rate of change of density of texture elements or edge elements in the image, based on the uniform density assumption. The geometric model is similar to that of Ohta et al. (1981) in Figure 9.3. If S^* be the area of a region on the surface and S the area of the corresponding region in the image, then we have the following equation from equations (9.5) and (9.8).

$$S^* = \lambda S \tag{9.11}$$

with

$$\lambda = \frac{c^2 \sqrt{p^2 + q^2 + 1}}{(1 - pX - qY)^3}$$

where (X, Y) is the center of gravity of the region in the image.

The uniform density assumtion states that if S_1^* and S_2^* are the areas of any two regions on the surface and they contain K_1 and K_2 texture elements (or edge elements), respectively, then $K_1/S_1^* = K_2/S_2^*$. Considering any two regions in the image with area S_1 and S_2 that contain K_1 and K_2 texture elements (or edge elements), respectively, then under the assumption that the elements on the surface are uniformly distributed and from equation (9.11), we have

$$\frac{K_1}{\lambda_1 S_1} = \frac{K_2}{\lambda_2 S_2} \tag{9.12}$$

with

$$\lambda_1 = \frac{c^2 \sqrt{1 + p^2 + q^2}}{(1 - pX_1 - qY_1)^3}$$

$$\lambda_2 = \frac{c^2 \sqrt{1 + p^2 + q^2}}{(1 - pX_2 - qY_2)^3}$$

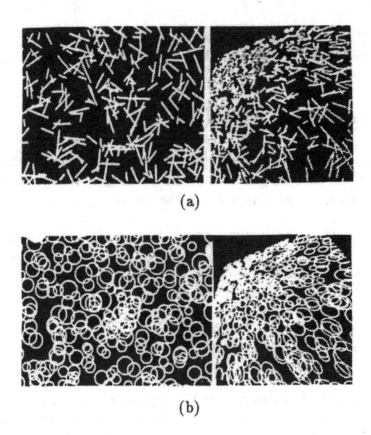

(a)

(b)

Figure 9.6: Examples of images of planes parallel to the image plane (left) and the images of the rotated planes (right) (from Aloimonos amd Swain, 1985).

where (X_1, Y_1) and (X_2, Y_2) are the centers of gravity of the two regions in the image, respectively. From this we have the following equation.

$$\left(\sqrt[3]{\frac{K_2}{K_1}\frac{S_1}{S_2}}X_2 - X_1 \right) p + \left(\sqrt[3]{\frac{K_2}{K_1}\frac{S_1}{S_2}}Y_2 - Y_1 \right) q = \sqrt[3]{\frac{K_2}{K_1}\frac{S_1}{S_2}} - 1 \quad (9.13)$$

The above equation represents a line in p-q space. Any two regions in the image constrain (p, q) to lie on a line in the gradient space. Thus, the coordinate of the intersecting point of lines given by two or more pairs of regions gives the surface gradient (p, q).

Figure 9.6 shows examples of images of planes parallel to the image plane and the images of the rotated planes. The orientations of the rotated planes have been estimated properly based on the assumption of uniform edge density.

9.2 Converging Lines

Converging lines in a perspective image constrain the orientation of a planar surface, assuming that they are parallel in 3D space, because they determine the vanishing point in the image.

9.2.1 Nakatani, Kimura, Saito, and Kitahashi (1980)

They used *Hough transformation* and inverse Hough transformation to get a set of converging lines in an image.

1) Edge Detection: Detect the edge element with direction θ at every point (x, y) in the image, and suppress weak edge elements.

2) Hough Transformation: Map the edge element into ρ-θ space which represents the line equation in the x-y space as $\rho = x \cos \theta + y \sin \theta$ and accumulate the numbers of edge elements on the same line. A large number of edge elements accumulated in the ρ-θ space implies the existence of a long line.

3) Inverse Hough Transformation: Map the numbers in the ρ-θ space back into the points on the corresponding lines in the x-y space.

4) Extraction of Vanishing Point: Define the point where the number is the maximum in the x-y space as the vanishing point.

For example, Figure 9.7 shows the representative lines passing through the vanishing point in the image. If the image is taken without rotation and the vanishing point is at (x_v, y_v) in the coordinates whose origin is at the center of the image, the slant angle (angle between the surface and the line of sight) is computed by $\theta = \tan^{-1}(y_v/f)$ where f is the focal length.

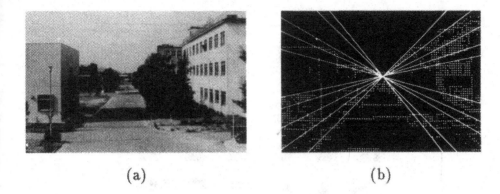

(a)   (b)

Figure 9.7: Extraction of a vanishing point: (a) Outdoor scene; (b) Lines passing through a vanishing point in the image (from Nakatani, Kimura, Sato, and Kitahashi, 1980).

9.2.2 Kender (1979a)

He presented an elegant way to estimate the vanishing line from more than two sets of converging lines in the image.

1) Detect the edge element with direction θ at every point (x, y) in the image.

2) Represent the edge element as the line $\rho = x \cos \theta + y \sin \theta$.

3) Transform the line into a point T in the p-q space by $T = (p, q) = ((f/\rho) \cos \theta, (f/\rho) \sin \theta)$ where f is the focal length. When we transform a set of lines that converges to a point in the image, the transformed points lie on a single line in the p-q space.

4) Detect straight lines in the p-q space (by Hough transformation).

5) Their intersection point is proved to be the gradient of the original surface.

For example, a synthetic image (a simulated building face) in Figure 9.8(a) is used to estimate the surface gradient. The edge elements detected in Figure 9.8(b) are transformed into the points in the p-q space,

149

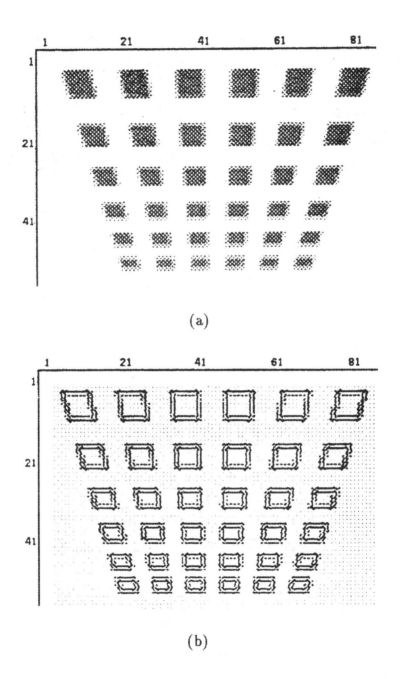

(a)

(b)

Figure 9.8: (a) A synthetic textured scene; (b) Edge elements; (from Kender, 1979a).

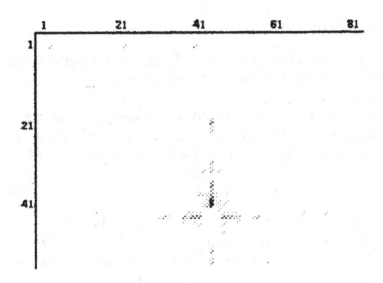

Figure 9.9: Transformed edge elements in the gradient space (from Kender, 1979a).

as shown in Figure 9.9. The coordinates $(0, -1)$ of the intersection of two lines represents the surface gradient.

9.3   Normalized Texture Property Map

When texture elements on a surface are known in advance, we can make the *normalized texture property map* to represent the apparent properties of texture in the image as a function of surface gradient (Kender, 1979b). The map is analogous to a reflectance map in shape-from-shading (Horn, 1977). The problem is underconstrained, and methods would be required to determine a solution. One method is the same as shape-from-shading which uses the portions of which the surface gradient are known and propagates them by *relaxation* or *regularization* based on smoothness assumption (Poggio et al., 1985). The other method is to use two or more independent properties of each texture elements, each characterized by its own normalized property map. This is analogous to photometric stereo (Woodham, 1980), and locally determines the surface gradient.

151

9.3.1 Kender (1979b)

He proposed to use such properties of texture elements as slope, length, area, density, angle between texture elements, eccentricity, skewed symmetry axes, etc.

For example, consider a vector $(L, 0)$ in the image. Its length is L. When it is a projection of the vector on the surface $-z = px + qy + c$, the real vector is $(L, 0, -pL)$. Then the real length is computed by

$$L_n = L\sqrt{1 + p^2} \qquad (9.14)$$

The contour form of the map is shown in Figure 9.10. The contours indicate those surface gradients which cause identical distortion if the real length is L_n. The result for any other slope is simply a rotation of the normalized map to that corresponding direction.

Another example is the map for an apparent density D. The real density is computed by

$$D_n = D\sqrt{1 + p^2 + q^2} \qquad (9.15)$$

The contour form of the map is rotationally symmetric.

9.3.2 Ikeuchi (1980)

He proposed a method to recover a curved surface from the spherical projection of texture elements on the surface. A viewer is assumed to be located at the center of a sphere. A point on the surface is projected on to the image sphere with respect to its center. Since the line of sight is perpendicular to the image sphere, the distortion of a pattern depends only on the direction of the line of sight and the surface normal.

Let a texture element be associated with two perpendicular axis vector. When the angle between the direction of the line of sight and the surface normal is ω, the magnitude of the cross product of the two axis vector projections is proportional to $\cos\omega$, and the sum of squares of their lengths is proportional to $1 + \cos^2\omega$. Both the cross product and the sum of squares depend on the distance. Then the ratio of the former to the latter only depends on the slant angle ω as follows.

$$I = \frac{\cos\omega}{1 + \cos\omega} \qquad (9.16)$$

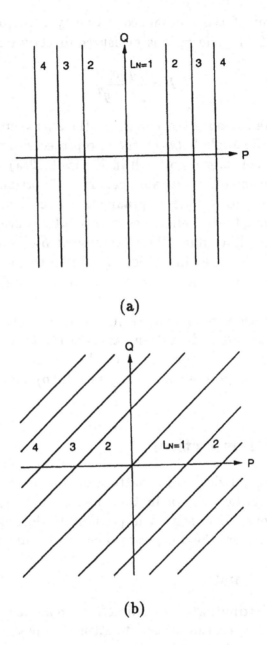

(a)

(b)

Figure 9.10: (a) The normalized textural property map for the unit horizontal length element; (b) The normalized textural property map for the unit length element which is oriented at 45 degrees (from Kender, 1979b).

The contour form of the normalized property map for the ratio I is shown in Figure 9.11. The ratio is measured directly from the image by

$$I = \frac{fg \sin \tau}{f^2 + g^2} \tag{9.17}$$

where f and g are the observed lengths of the axis vectors on the image sphere, and τ is the angle between the two projected axis vectors.

For example, consider a golf ball in Figure 9.12(a) which has many circular texture elements on its surface. The distortion of these circles are used to recover local surface gradients. The distortion value I is obtained by drawing two parallel lines in an arbitrary orientation tangent to each circle. The line connecting the tangent points on each circle is one of the axis vectors. A line which is parallel to the tangential lines through the center of the first axis vector, in the circle, is the second axis vector.

The iterative propagation algorithm determines the surface gradients using the map of the distortion values of the texture elements and assuming that the surface normals at the boundary of the ball are perpendicular to the image boundary. Figure 9.12(b) shows the recovered surface.

9.4 Shape Distortion

When the original shape of a pattern on the surface is known, the apparent shape of the pattern in the image can be given for all the surface orientations. Inversely, it is possible to determine the surface orientation locally from the observed shape of the pattern in the image.

9.4.1 Witkin (1982)

He used the distribution of edge directions in an image, which reflects the shape of patterns on the surface, to estimate the surface orientation. The form of the distribution depends on the orientation of the surface; the apparent shape is compressed in the direction of steepest inclination (the tilt direction). The amount of compression varies with the angle between the image plane and the surface (the slant angle).

154

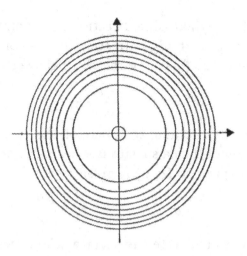

Figure 9.11: The normalized property map for the ratio of the cross product of the two axis vectors to the sum of the lengths of the two axis vectors (from Ikeuchi, 1980).

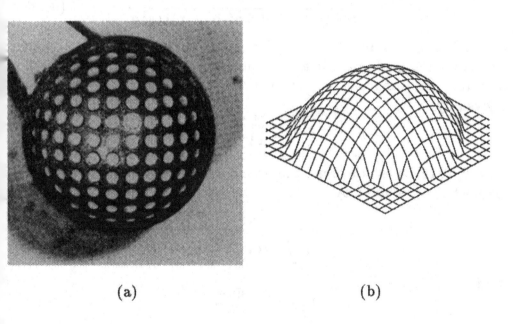

(a)   (b)

Figure 9.12: (a) A golf ball; (b) The recovered surface (from Ikeuchi 1980).

Let β denote the original edge direction at a point on the surface, and α^* denote the apparent edge direction at the corresponding point in the image. There is the following relation between α^* and β.

$$\alpha^* = \tan^{-1}\frac{\tan\beta}{\cos\sigma} + \tau \qquad (9.18)$$

where τ and σ are the tilt and the slant of the surface, respectively. When the distribution of β on a surface is known, the probability density function of α^* for a specified (σ, τ) is given by

$$P(\alpha^* \mid \sigma, \tau) = P(\beta \mid \sigma, \tau)\frac{\partial\beta}{\partial\alpha^*} \qquad (9.19)$$

The problem is how to estimate the surface orientation (σ, τ) from the observed distribution of α^*.

When the distribution of β is assumed to be homogeneous in all directions (isotropy), differentiating β with respect to α^* gives

$$\frac{\partial\beta}{\partial\alpha^*} = \frac{\cos\sigma}{\cos^2(\alpha^* - \tau) + \sin^2(\alpha^* - \tau)\cos^2\sigma} \qquad (9.20)$$

and $P(\beta \mid \sigma, \tau)$ is simply $1/\pi$. Let $A^* = \{\alpha_1^*, \ldots, \alpha_n^*\}$ denote a set of edge directions measured in the image. Assuming that the set of measures of edge direction is independent, the joint density function of A^* for (σ, τ) is given by

$$P(A^* \mid \sigma, \tau) = \prod_{i=1}^{n} P(\alpha_i^* \mid \sigma, \tau) \qquad (9.21)$$

According to Bayes' theorem, the probability density function of (σ, τ) for a set of α^* is given by

$$P(\sigma, \tau \mid A^*) = \frac{P(\sigma, \tau)P(A^* \mid \sigma, \tau)}{\displaystyle\int\int P(\sigma, \tau)P(A^* \mid \sigma, \tau)d\sigma d\tau} \qquad (9.22)$$

where

$$P(\sigma, \tau) = \frac{\sin\sigma}{\pi} \qquad (9.23)$$

The value of (σ, τ) for which this function gives a maximum is the maximum likelihood estimate for the surface gradient.

Figure 9.13 shows the result of surface estimation from the local distributions of edge directions in the image.

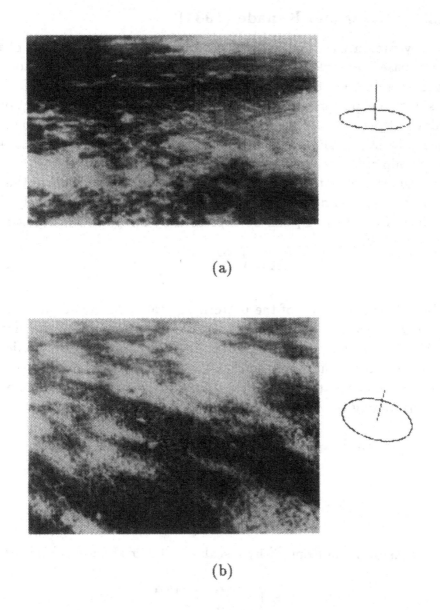

Figure 9.13: Surface orientation estimates based on the assumption of isotropy (the estimated surface orientation is indicated by an ellipse, representing the projected appearence a circle lying on the surface would have) (from Witkin, 1982).

9.4.2 Walker and Kanade (1984)

They presented a method of recovering the shape of a solid of revolution based on surface normals derived from the apparent distortions of patterns on the object. Their assummption is that a pair of affine-transformable patterns in the image are (orthographic) projection of similar patterns in the 3D space, i.e., they can be overlapped by scale change, rotation, and translation (Kanade and Kender, 1981). This relationship is illustrated in Figure 9.14.

Consider two patterns P_1 and P_2 in the image, and place the origin of the x–y coordinates at each center of gravity, respectively. The transform from P_2 to P_1 can be expressed by a regular 2×2 matrix

$$A = \begin{pmatrix} a_{11} & a_{12} \\ a_{21} & a_{22} \end{pmatrix} \tag{9.24}$$

Let P_i be projection of the pattern P_i' drawn on the 3D plane $-z = p_i x + q_i y$ for $i = 1, 2$. The frontal view of the pattern P_i' on the plane is obtained by rotating the coordinates first by ϕ_i around the y-axis and then by θ_i around the x-axis so that the plane is represented as $-z' = 0$ in the new coordinates for $i = 1, 2$, individually. We have the following relations among ϕ_i, θ_i, p_i, and q_i:

$$\sin \phi_i = \frac{p_i}{\sqrt{p_i^2 + 1}} \qquad \cos \phi_i = \frac{1}{\sqrt{p_i^2 + 1}}$$

$$\sin \theta_i = \frac{q_i}{\sqrt{p_i^2 + q_i^2 + 1}} \qquad \cos \theta_i = \frac{\sqrt{p_i^2 + 1}}{\sqrt{p_i^2 + q_i^2 + 1}} \tag{9.25}$$

P_1' is transformable from P_2' by a scalar factor σ and a rotation matrix

$$R = \begin{pmatrix} \cos \alpha & -\sin \alpha \\ \sin \alpha & \cos \alpha \end{pmatrix} \tag{9.26}$$

The 2D mapping from P_i' to P_i is represented by the following 2×2 matrix T_i which is a submatrix of the usual 3D rotation matrix.

$$T_i = \begin{pmatrix} \cos \phi_i & -\sin \phi_i \sin \theta_i \\ 0 & \cos \theta_i \end{pmatrix} \tag{9.27}$$

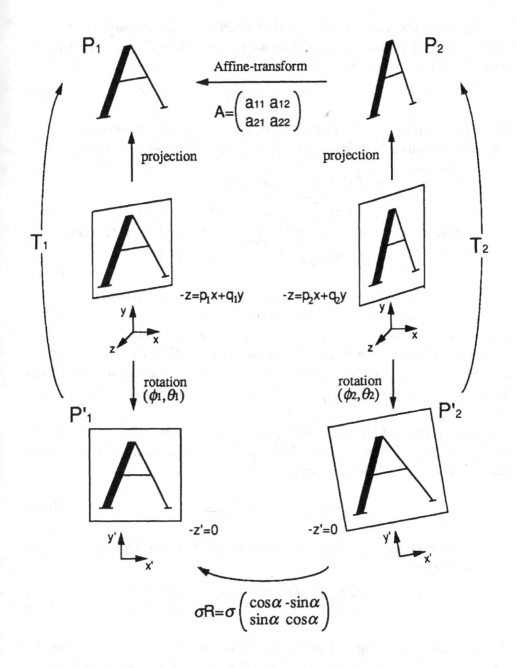

Figure 9.14: Relationship between Affine-transformable patterns (from Kanade and Kender, 1981).

Equating the two transforms that start from P_2' to reach at P_1: one following the diagram counter-clockwise $P_2' \to P_2 \to P_1$, the other clockwise $P_2' \to P_1' \to P_1$, gives the following equation:

$$AT_2 = T_1 \sigma R \tag{9.28}$$

By eliminating σ and α, and using p_i and q_i instead of $\sin \phi_i$ and $\cos \theta_i$, we have two equations in terms of p_i, q_i, and the elements of A.

$$\sqrt{p_2^2 + q_2^2 + 1}(a_{11}(p_1^2 + 1) + a_{21}p_1 q_1)$$
$$= \sqrt{p_1^2 + q_1^2 + 1}(a_{22}(p_2^2 + 1) + a_{21}p_2 q_2)$$
$$(-a_{12}(p_2^2 + 1) + a_{11}p_2 q_2)(p_1^2 + 1) - (a_{22}(p_2^2 + 1) + a_{21}p_2 q_2)p_1 q_1$$
$$= a_{21}\sqrt{p_1^2 + q_1^2 + 1}\sqrt{p_2^2 + q_2^2 + 1}$$

$$\tag{9.29}$$

As a result, the assumption of affine-transformable patterns yeilds a constraint on the surface gradient solely by the matrix A. The matrix is determined by the relation between P_2 and P_1 without knowing either the original patterns or their relationships in the 3D space. If (p_2, q_2) are known, then the solutions of the above equations for (p_1, q_1) are of the form (p_0, q_0) and $(-p_0, -q_0)$, symmetric about the origin of the gradient space.

Let Γ represent the slant of a pattern, the angle between the line of sight and the surface normal of the pattern. The slant of the pattern may be computed from its gradient:

$$\cos \Gamma = \frac{1}{\sqrt{p^2 + q^2 + 1}} \tag{9.30}$$

Then, the following relation is also derived from the equation 9.28 by eliminating α:

$$\frac{\det(A)}{\sigma^2} = \frac{\sqrt{p_2^2 + q_2^2 + 1}}{\sqrt{p_1^2 + q_1^2 + 1}} = \frac{\cos \Gamma_1}{\cos \Gamma_2} \tag{9.31}$$

As a result, if the original patterns are assumed to be all the same size ($\sigma = 1$), it is possible to sort the patterns by slant according to the determinants of the transform.

Consider the image of a solid of revolution with patterns of diamonds on it in Figure 9.15(a). Once the affine transforms have been found, a

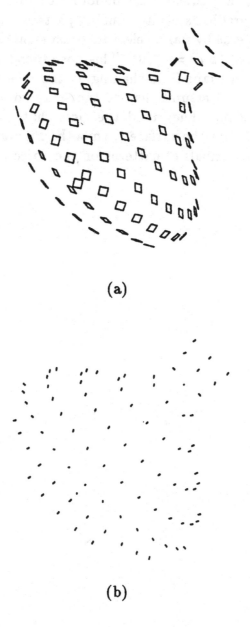

(a)

(b)

Figure 9.15: (a) Image of a solid of revolution with patterns of diamonds on it; (b) Estimated surface normals of patterns (from Walker and Kanade, 1984).

pattern with known gradient must be found to apply the equation 9.29 to constrain the gradients of the remaing patterns. One of the method is to sort the patterns by slant, select the least slanted pattern, and then give it the gradient $(p,q) = (0,0)$. This corresponds to using the most brightnest point in shape-from-shading as the reference of the surface gradient. Additional assumtions are necessary to uniquely determine the surface gradients. They used the facts that the object is a solid of revolution and that the surface is smooth. Figure 9.15(b) shows the estimated surface normals of patterns on the object.

References

Ahuja, N. (1982), Dot pattern processing using Voronoi neighborhood, *IEEE Trans.*, **PAMI-4**, 336-343.

Alloimonos, J. and Swain, J. (1985), Shape from texture, *Proc. 9th Int. Joint Conf. on Artificial Intelligence*, 926-931.

Bajcsy, R. (1973), Computer description of textured surfaces, *Proc. 3rd Int. Joint Conf. on Artificial Intelligence*, 572-579.

Bajcsy, R. and Lieberman L. (1976), Texture gradient as a depth cue, *Computer Graphics and Image Processing*, **5**, 52-67.

Blum, H. (1964), A transformation for extracting new descriptions of shape, *Symposium on Models for the Perception of Speech and Visual Form*, M.I.T. Press.

Brodats, P. (1966), *Textures*, New York: Dover, 1966.

Coleman, G. and Andrews, H. C. (1979), Image segmentation by clustering, *Proc. IEEE*, **67**, 773-785.

Conners, R. W. and Harlow, C. A. (1980), A theoretical comparison of texture algorithms, *IEEE Trans.*, **PAMI-2**, 204-222.

Davis, L. S., Rosenfeld, A., and Weszka, J. S. (1975), Region extraction by averaging and thresholding, *IEEE Trans.*, **SMC-5**, 383-388.

Davis, L. S., Johns, S. A., and Aggarwal, J. K. (1979), Texture analysis using generalized co-occurrence matrices, *IEEE Trans.*, **PAMI-1**, 251-259.

Davis, L. S. (1979), Computing the spatial structure of cellular textures, *Computer Graphics and Image Processing*, **11**, 111-122.

Deguchi, K. and Morishita, I. (1978), Texture characterization and texture-based partitioning using two-dimensional linear estimation, *IEEE Trans.*, **C-27**, 739-745.

Deutsch, E. S. and Belknap, N. J. (1972), Texture description using neighborhood information, *Computer Graphics and Image Processing*, **1**, 145-168.

Duda, R. O. and Hart, P. E. (1973), *Pattern Classification and Scene Analysis*, New York: Wiley.

Dyer, C. R. and Rosenfeld, A. (1976), Fourier texture features: Suppression of aperture effects, *IEEE Trans.*, **SMC-6**, 703-705.

Ehrick, R. W. and Foith, J. P. (1978), A view of texture topology and texture description, *Computer Graphics and Image Processing*, **8**, 174-202.

Eklundh, J. O. (1979), On the use of Fourier phase features for texture discrimination, *Computer Graphics and Image Processing*, **9**, 199-201.

Faugeras, J. O. (1978), Texture analysis and classification using a human visual model, *Proc. 4th Int. Conf. Pattern Recognition*, 549-552.

Funakubo, N. (1978), Analysis and discussion osf a coarseness detector, *IEEE Trans.*, **SMC-8**, 229-232.

Gagalowicz, A. (1978), Analysis of texture using stochastic model, *Proc. 4th Int. Conf. Pattern Recognition*, 541-544.

Gagalowicz, A. (1979), Stochastic texture fields synthesis from a priori given second order statistics, *Proc. Pattern Recognition and Image Processing*, 376-381.

Galloway, M. M. (1975), Texture classification using gray level run length, *Computer Graphics and Image Processing*, **4**, 172-179.

Gibson, J. J. (1950), *The Perception of the Visual World*, Houghton Mifflin, Boston.

Haralick, R. M., Shanmugam, K., and Dinstein, I. (1973), Textural features for image classification, *IEEE Trans.*, **SMC-3**, 610-621.

Haralick, R. M. (1978), Statistical and structural approach to texture, *Proc. 4th Int. Joint Conf. Pattern Recognition*, 45-69.

Hassher, M. and Sklansky, J. (1978), Marcov random field models of digitized image texture, *Proc. 4th Int. Joint Conf. Pattern Recognition*, 538-540.

Hayes, K. C., Shah, A.N., and Rosenfeld, A. (1974), Texture coarseness: Further experiments, *IEEE Trans.*, **SMC-4**, 467-472.

Hilditch, J. (1969), Linear skeleton from square cupboards, *Machine Intelligence*, 6, Edinbargh Univ. Press, 403-420.

Horn, B. K. P. (1977), Understanding image intensities, *Artificiall Intelligence*, **8**, 201-231.

Ikeuchi, K. (1980), Shape from regular-patterns, *Proc. 5th Int. Conf. on Pattern Recognition*, 1032-1039.

Julesz, B. (1975), Experiments in the visual perception of texture, *Scientific American*, **232**, 34-43.

Julesz, B. and Caelli, T. (1979), On the limits of Fourier decompositions in visual texture perception, *Perception*, **8**, 69-73.

Kanade, T. and Kender, J. R. (1980), Mapping image properties into shape constraints: skewed symmetry, affine-transformable patterns, and the shape-from-texture paradigm, *Proc. IEEE Workshop on Picture Data Description and Management*, 130-135, Asilmoar, CA.

Kender, J. R. (1979a), Shape from texture: An aggregation transform that maps a class of textures into surface orientation, *Proc. 6th Int. Joint Conf. on Artificial Intelligence*, 475-480.

Kender, J. R. (1979b), Shape from texture: A computational paradigm, *Proc. Image Understanding Workshop*, 134-138.

Lipkin, B. C. and Rosenfeld, A., Eds.(1970), *Picture Processing and Psychopictorics*, Academic Press, New York.

Mack, G. A., Jain, V. K., and Eikman, E. A. (1978), Polar reference texture in radiocoloid liver image, *Proc. Pattern Recognition and Image Processing*, 369-371.

Maleson, J. T., Brown, C. M., and Feldman, J. A. (1977), Understanding natural texture, *Proc. Image Understanding Workshop*, 19-27.

Mandelbrot, B. B. (1982), *The fractal Geometry of Nature*, W. H. Freeman, San Francisco.

Marr, D. (1976), Early processing of visual information, *Philosophical Transactions of the Royal Society of London*, Series B, **275**, 483-524.

Marr, D. and Hildreth, E. (1980), Theory of edge detection, *Proc. the Royal Society of London*, Series B, **207**, 187-217.

Matsuyama, T. (1986), Expert systems for image processing: An overview, in *Signal Processing* III (Young, I. T. et al., Eds.), Elsevier, New York, 869-872.

McCormic, B. H. and Jayaramamurthy, S. N. (1974), Time series model for texture synthesis, *Computer and Information Science*, 4, 329-343.

Mitchell, O. R., Myer, C. R., and Boyne, W. (1977), A max-min measure for image texture analysis, *IEEE Trans.*, **C-26**, 408-414.

Muerle, J. L. (1970), Some thoughts on texture discrimination by computer, in *Picture Processing and Psychopictorics* (Lipkin, B. C. and Rosenfeld, A., Eds.), Academic Press, New York, 371-379.

Nagao, M., Tanabe, H., and Ito, K. (1976), Agricultural land use classification of aerial photographs by histogram similarity method, *Proc. 3rd Int. Joint Conf. Pattern Recognition*, 669-672.

Nagao, M. and Matsuyama, T. (1979), Edge preserving smoothing, *Computer Graphics and Image Processing*, **9**, 394-407.

Nakatani, H., Kimura, S., Saito, O., and Kitahashi, T. (1980), Extraction of vanishing point and its application, *Proc. 5th Int. Conf. Pattern Recognition,* 370-372.

Nevatia, R., Price, K., and Vilnrotter, R. (1979), Describing natural textures, *Proc. 6th Int. Joint Conf. Artificial Intelligence,* 642-644.

Ohlander, R., Price, K., and Reddy, D. R. (1979), Picture segmentation using a recursive splitting method, *Computer Graphics and Image Processing,* **8**, 313-333.

Ohta, Y., Maenobu, K., and Sakai, T. (1981), Obtaining surface orientation from texels under perspective projection, *Proc. 7th Int. Joint Conf. on Artificial Intelligence,* 746-751.

Pavlidis, T. (1978), A review of algorithms for shape analysis, *Computer Graphics and Image Processing,* **7**, 243-258.

Pentland, A. P. (1984), Fractal-based description of natural scenes, *IEEE Trans.,* **5**, 661-674.

Pickett, R. M. (1970), Visual analysis of texture in the detection and recognition of objects, in *Picture Processing and Psychopictorics* (Lipkin, B. C. and Rosenfeld, A., Eds.), Academic Press, New York, 298-308.

Poggio, T., Torre, V. and Koch, C. (1985), Computational vision and regularization theory, *Nature,* **317**, 314-319.

Pratt, W. K. and Faugeras, O. C. (1978), Development and evaluation of stochastic-based visual texture features, *IEEE Trans.,* **SMC-8**, 796-804.

Ramsey, R. G. (1981), *Neuroradiology with Computed Tomography,* Saunders, Philadelphia.

Read, J. S. and Jayaramamurthy, S. N. (1972), Automatic generation of texture feature detectors, *IEEE Trans.,* **C-21**, 803-811.

Rosenfeld, A. and Pfaltz, J. L. (1968), Distance functions on digital pictures, *Pattern Recognition,* **1**, 33-61.

167

Rosenfeld, A. and Lipkin, B. S. (1970), Texture synthesis, in *Picture Processing and Psychopictorics* (Lipkin, B. C. and Rosenfeld, A., Eds.), Academic Press, New York, 309-345.

Rosenfeld, A. and Thurston M. (1971), Edge and curve detection for visual scene analysis, *IEEE Trans.*, **C-20**, 562-569.

Rosenfeld, A., Thurston M., and Lee, Y. H. (1972), Edge and curve detection: Further experiments, *IEEE Trans.*, **C-21**, 677-715.

Rosenfeld, A. (1975), A note on automatic detection of texture gradients, *IEEE Trans.*, **C-24**, 988-991.

Schachter, G. J., Rosenfeld, A., and Davis, L. S. (1978), Random mosaic models for textures, *IEEE Trans.*, **SMC-8**, 694-702.

Schatz, B. R. (1977), Computation of immediate texture discrimination, *Proc. 5th Int. Joint Conf. on Artificial Intelligence*, p. 708.

Strong III, J. P. and Rosenfeld, A. (1973), A region coloring technique for scene analysis, *Comm. ACM*, **16**, 237-246.

Sutton, R. N. and Hall, E. L. (1972), Texture measurement for automatic classification of pulmonary disease, *IEEE Trans.*, **C-21**, 667-676.

Tamura, H., Mori, S., and Yamawaki, T. (1978), Textural features corresponding to visual perception, *IEEE Trans.*, **SMC-8**, 460-473.

Tamura, H., Sakane, S., Tomita, F., Yokoya, N., Kaneko, N., and Sakaue, K. (1983), Design and implementation of SPIDER–A transportable image processing software package, *Computer Vision, Graphics, and Image Processing*, **23**, 273-294.

Thompson, W. B. (1977), Texture boundary analysis, *IEEE Trans.*, **C-26**, 272-276.

Tomita, F., Yachida, M., and Tsuji, S. (1973), Detection of homogeneous regions by structural analysis, *Proc. 3rd Int. Joint Conf. Artificial Intelligence*, 564-571.

Tomita, F. and Tsuji, S. (1977), Extraction of multiple regions by smoothing in selected neighborhoods, *IEEE Trans.*, **SMC-7**, 107-109.

Tomita, F. (1981), Hierarchical description of textures, *Proc. 7th Int. Joint Conf. on Artificial Intelligence*, 728-733.

Tomita, F., Shirai, Y., and Tsuji, S. (1982), Description of textures by a structural analysis, *IEEE Trans.*, **PAMI-4**, 183-191.

Tomita, F. (1983), A learning vision system for 2D object recognition, *Proc. 8th Int. Joint Conf. on Artificial Intelligence*, 1132-1135.

Tomita, F. (1988), Interactive and automatic image recognition system, *Machine Vision and Application*, **1**, 59-69.

Tou, J. T., Kao, D. B., and Chang, Y. S. (1976), Pictorial texture analysis and synthesis, *Proc. 3rd Int. Joint Conf. on Pattern Recognition*, 590-590P.

Tou, J. T. and Chang, Y. S. (1976), Picture understanding by machine via textural feature extraction, *Proc. Pattern Recognition and Image Processing*, 392-399.

Tsuji, S. and Tomita, F. (1973), A structural analyzer for a class of textures, *Computer Graphics and Image Processing*, **2**, 216-231.

Walker, E. L. and Kanade, T. (1984), Shape recovery of a solid of revolution from apparent distortions of patterns, *Carnegie-Mellon Univ. Tech. Rep. CMU-CS-80-133*.

Weszka, J. S., Dyer, C. R., and Rosenfeld, A. (1976a), A comparative study of texture measures for terrain classification, *IEEE Trans.*, **SMC-6**, 269-285.

Weszka, J. S. and Rosenfeld, A. (1976b), An application of texture analysis to material inspection, *Pattern Recognition*, **8**, 195-200.

Winston, P. H. (1975), Learning structure descriptions from examples, in *The Psychology of Computer Vision* (Winston, P. H., Ed.) McGraw Hill, New York, 157-210.

Winston, P. H. (1977), *Artificiall Intelligence*, Addison Wesley.

Witkin, A. P. (1981), Recovering surface shape and orientation from texture, *Artificial Intelligence*, **17**, 17-46.

Woodham, R. J. (1981), Analyzing images of curved surfaces, *Artificial Intelligence*, **17**, 117-140.

Zahn, C. T. and Roskies, P. Z. (1972), Fourier descriptors for plane closed curves, *IEEE Trans.*, **C-21**, 269-281.

Zobrist, A. L. and Thompson, W. B. (1975), Building a distance function for gestalt grouping, *IEEE Trans.*, **C-24**, 718-728.

Zucker, S. W., Rosenfeld, A., and Davis, L. S. (1975), Picture segmentation by texture discrimination, *IEEE Trans.*, **C-24**, 718-728.

Zucker, S. W. (1976), Toward a model of texture, *Computer Graphics and Image Processing*, **5**, 190-202.

Index